AF886303

SOTHIS PRESS

Sothis Press aims to completely reimagine the public domain book with an emphasis on universal access to captivatingly designed print editions. We believe that technology will enable every single book ever published to be perpetually accessible in beautiful print form. Sothis is beauty in ubiquity.

www.sothis.press
1 Waterhouse Square, London, ECIN 2ST

Complete Examination Questions and Answers for Marine and Stationary Engineers

A Complete Engine Operator's Catechism

Calvin F. Swingle

SOTHIS PRESS

ISBN 978-1-0365-1279-8

Complete Examination Questions and Answers for Marine and Stationary Engineers is in the public domain.

This edition is published by Sothis Press The company is registered in England and Wales. The Company Number is 14900389.

Yours Fraternally
Calvin F Swingle

COMPLETE

Examination Questions and Answers

FOR

Marine and Stationary Engineers

A Complete Engine Operator's Catechism

Giving the Latest and Most Approved Answers to all Leading Questions which will be Asked for the Purpose of Examining and Determining the Qualifications of Applicants for Licenses for Engineers and for Persons having Charge of Steam Boilers as Approved by all Municipalities and Government Boards of Examining Engineers, both Stationary and Marine.

BY

CALVIN F. SWINGLE, M. E.

Author of "Twentieth Century Hand-Book for Steam Engineers and Electricians," "Modern Locomotive Engineering Hand-Book," "Steam Boilers: Their Construction, Care and Operation," Etc.

ILLUSTRATED

CHICAGO
FREDERICK J. DRAKE & CO., PUBLISHERS
1906

INTRODUCTION

The development of the science of steam engineering and the continually increasing demand for more power for manufacturing purposes and for transportation, both on land and sea, have in these modern times resulted in the creation of power plants, which are truly marvelous in their details when compared with the steam machinery of forty years ago. Even the last twenty years have witnessed tremendous developments along these lines, and we may imagine the effect it would have upon an engineer, who twenty years ago was counted as first class in his business, but who, having taken a Rip Van Winkle sleep of twenty years, is suddenly awakened and finds himself set down in the engine room of a first-class ocean steamer, or in the midst of one of our modern up-to-date power plants. The facts are, he would have hard work to recognize his surroundings. Even the steam gauges would indicate a pressure of 150 to 175 pounds more per square inch than did the old-time gauges. Therefore, in view of the remarkable improvements in steam machinery which have been made and are continually being made, it certainly behooves engineers to do their utmost to keep step with the march of progress. The author has endeavored, in the following pages, to place before his readers information in a catechetical form which will be found to cover all of the various details appertaining to the operation of modern steam plants, both stationary and marine.

C. F. S.

CHAPTER I

STEAM, HEAT, COMBUSTION, AND FUELS

Ques. 1.—What is steam?

Ans.—Steam is vapor of water.

Ques. 2.—At what temperature will water evaporate (boil) in the open air at sea level?

Ans.—212 degrees Fahrenheit.

Ques. 3.—If 1 cubic foot of water is evaporated at 212 degrees into steam at atmospheric pressure, how many cubic feet of steam will there be? In other words, what will the volume of the steam be?

Ans.—1,646 cubic feet.

Ques. 4.—Then what is the relative volume of steam at atmospheric pressure, and the water from which it was evaporated at 212 degrees?

Ans.—1,646 to 1.

Ques. 5.—What is the relative volume of steam at 200 pounds gauge pressure, and the water from which it was generated?

Ans.—132 to 1.

Ques. 6.—What is meant by the terms atmospheric pressure, gauge pressure, and absolute pressure, as applied to steam and other gases?

Ans.—The pressure in pounds exerted by the steam, or gas, on each square inch of the interior surface of the containing vessel, tending to rupture it.

Ques. 7.—What is vacuum?

Ans.—The absence of all pressure in the interior of a vessel.

Table 1, which follows, shows the physical properties of saturated steam from a perfect vacuum up to 1,000 pounds absolute pressure. It will be found convenient for reference.

TABLE I

PROPERTIES OF SATURATED STEAM

Vacuum Inches of Mercury	Absolute Pressure Lbs. per Sq. Inch	Temp. Degrees F.	Total Heat above 32° F. In the Water h Heat-units	Total Heat above 32° F. In the Steam H Heat-units	Latent Heat H-h Heat units	Relative Volume	Cubic Feet in 1 Lb. Wt. of Steam	Wt. of 1 Cubic Foot of Steam, Lbs.
29.74	.089	32.	0.	1091.7	1091.7	208,080	3333.3	.0003
29.67	.122	40.	8.	1094.1	1086.1	154,330	2472.2	.0004
29.56	.176	50.	18.	1097.2	1079.2	107,630	1724.1	.0006
29.40	.254	60.	28.01	1100.2	1072.2	76,370	1223.4	.0008
29.19	.359	70.	38.02	1103.3	1065.3	54,660	875.61	.0011
28.90	.502	80.	48.04	1106.3	1058.3	39,690	635.80	.0016
28.51	.692	90.	58.06	1109.4	1051.3	29,290	469.20	.0021
28.00	.943	100.	68.08	1112.4	1044.4	21,830	349.70	.0028
27.88	1.	102.1	70.09	1113.1	1043.0	20,623	334.23	.0030
25.85	2.	126.3	94.44	1120.5	1026.0	10,730	175.23	.0058
23.83	3.	141.6	109.9	1125.1	1015.3	7,325	118.00	.0085
21.78	4.	153.1	121.4	1128.6	1007.2	5,588	89.80	.0111
19.74	5.	162.3	130.7	1131.4	1000.7	4,530	72.50	.0137
17.70	6.	170.1	138.6	1133.8	995.2	3,816	61.10	.0163
15.67	7.	176.9	145.4	1135.9	990.5	3,302	53.00	.0189
13.63	8.	182.9	151.5	1137.7	986.2	2,912	46.60	.0214
11.60	9.	188.3	156.9	1139.4	982.4	2,607	41.82	.0239
9.56	10.	193.2	161.9	1140.9	979.0	2,361	37.80	.0264
7.52	11.	197.8	166.5	1142.3	975.8	2,159	34.61	.0289
5.49	12.	202.0	170.7	1143.5	972.8	1,990	31.90	.0314
3.45	13.	205.9	174.7	1144.7	970.0	1,846	29.60	.0338
1.41	14.	209.6	178.4	1145.9	967.4	1,721	27.50	.0363
0.00	14.7	212.0	180.9	1146.6	965.7	1,646	26.36	.0379

TABLE 1—*Continued*

Gauge Pressure Lbs. per Sq. In.	Absolute Pressure Lbs. per Sq. In.	Temp. Degrees F.	Total Heat Above 32° F.		Latent Heat H-h Heat-units	Relative Volume	Cubic Feet in 1 Lb. Wt. of Steam	Wt. of 1 Cubic Foot of Steam, Lbs.
			In the Water h Heat-units	In the Steam H Heat-units				
0.3	15	213.3	181.9	1146.9	965.0	1,614	25.90	.0387
1.3	16	216.3	185.3	1147.9	962.7	1,519	24.33	.0411
2.3	17	219.4	188.4	1148.9	960.5	1,434	23.00	.0435
3.3	18	222.4	191.4	1149.8	958.3	1,359	21.80	.0459
4.3	19	225.2	194.3	1150.6	956.3	1,292	20.70	.0483
5.3	20	227.9	197.0	1151.5	954.4	1,231	19.72	.0507
6.3	21	230.5	199.7	1152.2	952.6	1,176	18.84	.0531
7.3	22	233.0	202.2	1153.0	950.8	1,126	18.03	.0555
8.3	23	235.4	204.7	1153.7	949.1	1,080	17.30	.0578
9.3	24	237.8	207.0	1154.5	947.4	1,038	16.62	.0602
10.3	25	240.0	209.3	1155.1	945.8	998	16.00	.0625
11.3	26	242.2	211.5	1155.8	944.3	962	15.42	.0649
12.3	27	244.3	213.7	1156.4	942.8	929	14.90	.0672
13.3	28	246.3	215.7	1157.1	941.3	898	14.40	.0696
14.3	29	248.3	217.8	1157.7	939.9	869	13.91	.0719
15.3	30	250.2	219.7	1158.3	938.9	841	13.50	.0742
16.3	31	252.1	221.6	1158.8	937.2	816	13.07	.0765
17.3	32	254.0	223.5	1159.4	935.9	792	12.68	.0788
18.3	33	255.7	225.3	1159.9	934.6	769	12.32	.0812
19.3	34	257.5	227.1	1160.5	933.4	748	12.00	.0835
20.3	35	259.2	228.8	1161.0	932.2	728	11.66	.0858
21.3	36	260.8	230.5	1161.5	931.0	709	11.36	.0880
22.3	37	262.5	232.1	1162.0	929.8	691	11.07	.0903
23.3	38	264.0	233.8	1162.5	928.7	674	10.80	.0926
24.3	39	265.6	235.4	1162.9	927.6	658	10.53	.0949
25.3	40	267.1	236.9	1163.4	926.5	642	10.28	.0972
26.3	41	268.6	238.5	1163.9	925.4	627	10.05	.0995
27.3	42	270.1	240.0	1164.3	924.4	613	9.83	.1018
28.3	43	271.5	241.4	1164.7	923.3	600	9.61	.1040
29.3	44	272.9	242.9	1165.2	922.3	587	9.41	.1063
30.3	45	274.3	244.3	1165.6	921.3	575	9.21	.1086
31.3	46	275.7	245.7	1166.0	920.4	563	9.02	.1108
32.3	47	277.0	247.0	1166.4	919.4	552	8.84	.1131
33.3	48	278.3	248.4	1166.8	918.5	541	8.67	.1153
34.3	49	279.6	249.7	1167.2	917.5	531	8.50	.1176
35.3	50	280.9	251.0	1167.6	916.6	520	8.34	.1198
36.3	51	282.1	252.2	1168.0	915.7	511	8.19	.1221
37.3	52	283.3	253.5	1168.4	914.9	502	8.04	.1243

TABLE I—*Continued*

Gauge Pressure Lbs. per Sq. In.	Absolute Pressure Lbs. per Sq. In.	Temp. Degrees F.	Total Heat above 32° F. In the Water h Heat-units	Total Heat above 32° F. In the Steam H Heat-units	Latent Heat H-h Heat-units	Relative Volume	Cubic Feet in 1 Lb. Wt. of Steam	Wt. of 1 Cubic Foot of Steam, Lbs.
38.3	53	284.5	254.7	1168.7	914.0	492	7.90	.1266
39.3	54	285.7	256.0	1169.1	913.1	484	7.76	.1288
40.3	55	286.9	257.2	1169.4	912.3	476	7.63	.1311
41.3	56	288.1	258.3	1169.8	911.5	468	7.50	.1333
42.3	57	289.1	259.5	1170.1	910.6	460	7.38	.1355
43.3	58	290.3	260.7	1170.5	909.8	453	7.26	.1377
44.3	59	291.4	261.8	1170.8	909.0	446	7.14	.1400
45.3	60	292.5	262.9	1171.2	908.2	439	7.03	.1422
46.3	61	293.6	264.0	1171.5	907.5	432	6.92	.1444
47.3	62	294.7	265.1	1171.8	906.7	425	6.82	.1465
48.3	63	295.7	266.2	1172.1	905.9	419	6.72	.1488
49.3	64	296.8	267.2	1172.4	905.2	413	6.62	.1511
50.3	65	297.8	268.3	1172.8	904.5	407	6.53	.1533
51.3	66	298.8	269.3	1173.1	903.7	401	6.43	.1555
52.3	67	299.8	270.4	1173.4	903.0	395	6.34	.1577
53.3	68	300.8	271.4	1173.7	902.3	390	6.25	.1599
54.3	69	301.8	272.4	1174.0	901.6	384	6.17	.1621
55.3	70	302.7	273.4	1174.3	900.9	379	6.09	.1643
56.3	71	303.7	274.4	1174.6	900.2	374	6.01	.1665
57.3	72	304.6	275.3	1174.8	899.5	369	5.93	.1687
58.3	73	305.6	276.3	1175.1	898.9	365	5.85	.1709
59.3	74	306.5	277.2	1175.4	898.2	360	5.78	.1731
60.3	75	307.4	278.2	1175.7	897.5	356	5.71	.1753
61.3	76	308.3	279.1	1176.0	896.9	351	5.63	.1775
62.3	77	309.2	280.0	1176.2	896.2	347	5.57	.1797
63.3	78	310.1	280.9	1176.5	895.6	343	5.50	.1819
64.3	79	310.9	281.8	1176.8	895.0	339	5.43	.1840
65.3	80	311.8	282.7	1177.0	894.3	334	5.37	.1862
66.3	81	312.7	283.6	1177.3	893.7	331	5.31	.1884
67.3	82	313.5	284.5	1177.6	893.1	327	5.25	.1906
68.3	83	314.4	285.3	1177.8	892.5	323	5.18	.1928
69.3	84	315.2	286.2	1178.1	891.9	320	5.13	.1950
70.3	85	316.0	287.0	1178.3	891.3	316	5.07	.1971
71.3	86	316.8	287.9	1178.6	890.7	313	5.02	.1993
72.3	87	317.7	288.7	1178.8	890.1	309	4.96	.2015
73.3	88	318.5	289.5	1179.1	889.5	306	4.91	.2036
74.3	89	319.3	290.4	1179.3	888.9	303	4.86	.2058
75.3	90	320.0	291.2	1179.6	888.4	299	4.81	.2080

TABLE I—*Continued*

Gauge Pressure Lbs. per Sq. In.	Absolute Pressure Lbs. per Sq. In.	Temp. Degrees F.	Total Heat above 32° F. In the Water h Heat-units	Total Heat above 32° F. In the Steam H Heat-units	Latent Heat H-h Heat-units	Relative Volume	Cubic Feet in 1 Lb. Wt. of Steam	Wt. of 1 Cubic Foot of Steam, Lbs.
76.3	91	320.8	292.0	1179.8	887.8	296	4.76	.2102
77.3	92	321.6	292.8	1180.0	887.2	293	4.71	.2123
78.3	93	322.4	293.6	1180.3	886.7	290	4.66	.2145
79.3	94	323.1	294.4	1180.5	886.1	287	4.62	.2166
80.3	95	323.9	295.1	1180.7	885.6	285	4.57	.2188
81.3	96	324.6	295.9	1181.0	885.0	282	4.53	.2210
82.3	97	325.4	296.7	1181.2	884.5	279	4.48	.2231
83.3	98	326.1	297.4	1181.4	884.0	276	4.44	.2253
84.3	99	326.8	298.2	1181.6	883.4	274	4.40	.2274
85.3	100	327.6	298.9	1181.8	882.9	271	4.36	.2296
86.3	101	328.3	299.7	1182.1	882.4	268	4.32	.2317
87.3	102	329.0	300.4	1182.3	881.9	266	4.28	.2339
88.3	103	329.7	301.1	1182.5	881.4	264	4.24	.2360
89.3	104	330.4	301.9	1182.7	880.8	261	4.20	.2382
90.3	105	331.1	302.6	1182.9	880.3	259	4.16	.2403
91.3	106	331.8	303.3	1183.1	879.8	257	4.12	.2425
92.3	107	332.5	304.0	1183.4	879.3	254	4.09	.2446
93.3	108	333.2	304.7	1183.6	878.8	252	4.05	.2467
94.3	109	333.9	305.4	1183.8	878.3	250	4.02	.2489
95.3	110	334.5	306.1	1184.0	877.9	248	3.98	.2510
96.3	111	335.2	306.8	1184.2	877.4	246	3.95	.2531
97.3	112	335.9	307.5	1184.4	876.9	244	3.92	.2553
98.3	113	336.5	308.2	1184.6	876.4	242	3.88	.2574
99.3	114	337.2	308.8	1184.8	875.9	240	3.85	.2596
100.3	115	337.8	309.[illegible]	1185.0	875.5	238	3.82	.2617
101.3	116	338.5	310.2	1185.2	875.0	236	3.79	.2638
102.3	117	339.1	310.8	1185.4	874.5	234	3.76	.2660
103.3	118	339.7	311.5	1185.6	874.1	232	3.73	.2681
104.3	119	340.4	312.1	1185.8	873.6	230	3.70	.2703
105.3	120	341.0	312.8	1185.9	873.2	228	3.67	.2764
106.3	121	341.6	313.4	1186.1	872.7	227	3.64	.2745
107.3	122	342.2	314.1	1186.3	872.3	225	3.62	.2766
108.3	123	342.9	314.7	1186.5	871.8	223	3.59	.2788
109.3	124	343.5	315.3	1186.7	871.4	221	3.56	.2809
110.3	125	344.1	316.0	1186.9	870.9	220	3.53	.2830
111.3	126	344.7	316.6	1187.1	870.5	218	3.51	2851
112.3	127	345.3	317.2	1187.3	870.0	216	3.48	.2872
113.3	128	345.9	317.8	1187.4	869.6	215	3.46	.2894

TABLE I —*Contiuued*

Gauge Pressure Lbs. per Sq. In.	Absolute Pressure Lbs. per Sq. In.	Temp. Degrees F.	Total Heat Above 32° F.		Latent Heat H-h Heat-units	Relative Volume	Cubic Feet in 1 Lb. Wt. of Steam	Wt. of 1 Cubic Foot of Steam, Lbs.
			In the Water h Heat-units	In the Steam H Heat-units				
114.3	129	346.5	318.4	1187.6	869.2	213	3.43	.2915
115.3	130	347.1	319.1	1187.8	868.7	212	3.41	.2936
116.3	131	347.6	319.7	1188.0	868.3	210	3.38	.2957
117.3	132	348.2	320.3	1188.2	867.9	209	3.36	.2978
118.3	133	348.8	320.8	1188.3	867.5	207	3.33	.3000
119.3	134	349.4	321.5	1188.5	867.0	206	3.31	.3021
120.3	135	350.0	322.1	1188.7	866.6	204	3.29	.3042
121.3	136	350.5	322.6	1188.9	866.2	203	3.27	.3063
122.3	137	351.1	323.2	1189.0	865.8	201	3.24	.3084
123.3	138	351.8	323.8	1189.2	865.4	200	3.22	.3105
124.3	139	352.2	324.4	1189.4	865.0	199	3.20	.3126
125.3	140	352.8	325.0	1189.5	864.6	197	3.18	.3147
126.3	141	353.3	325.5	1189.7	864.2	196	3.16	.3169
127.3	142	353.9	326.1	1189.9	863.8	195	3.14	.3190
128.3	143	354.4	326.7	1190.0	863.4	193	3.11	.3211
129.3	144	355.0	327.2	1190.2	863.0	192	3.09	.3232
130.3	145	355.5	327.8	1190.4	862.6	191	3.07	.3253
131.3	146	356.0	328.4	1190.5	862.2	190	3.05	.3274
133.3	148	357.1	329.5	1190.9	861.4	187	3.02	.3316
135.3	150	358.2	330.6	1191.2	860.6	185	2.98	.3358
140.3	155	360.7	333.2	1192.0	858.7	179	2.89	.3463
145.3	160	363.3	335.9	1192.7	856.9	174	2.80	.3567
150.3	165	365.7	338.4	1193.5	855.1	169	2.72	.3671
155.3	170	368.2	340.9	1194.2	853.3	164	2.65	.3775
160.3	175	370.5	343.4	1194.9	851.6	160	2.58	.3879
165.3	180	372.8	345.8	1195.7	849.9	156	2.51	.3983
170.3	185	375.1	348.1	1196.3	848.2	152	2.45	.4087
175.3	190	377.3	350.4	1197.0	846.6	148	2.39	.4191
180.3	195	379.5	352.7	1197.7	845.0	144	2.33	.4296
185.3	200	381.6	354.9	1198.3	843.4	141	2.27	.4400
190.3	205	383.7	357.1	1199.0	841.9	138	2.22	.4503
195.3	210	385.7	359.2	1199.6	840.4	135	2.17	.4605
200.3	215	387.7	361.3	1200.2	838.9	132	2.12	.4707
205.3	220	389.7	362.2	1200.8	838.6	129	2.06	.4852
245.3	260	404.4	377.4	1205.3	827.9	110	1.76	.5686
285.3	300	417.4	390.9	1209.2	818.3	96	1.53	.6515
485.3	500	467.4	443.5	1224.5	781.0	59	.94	1.062
685.3	700	504.1	482.4	1235.7	753.3	42	.68	1.470
985.3	1000	546.8	528.3	1248.7	720.3	30	.48	2.082

Ques. 8.—How much pressure does the atmosphere exert upon the surface of the earth?

Ans.—14.7 pounds upon each square inch of the earth's surface.

Ques. 9.—What is understood by gauge pressure?

Ans.—Gauge pressure is the pressure over and above the 14.7 pounds atmospheric pressure.

Ques. 10.—What is absolute pressure?

Ans.—Absolute pressure is the total pressure above a perfect vacuum. It equals the sum of the gauge pressure and the atmospheric pressure.

Ques. 11.—How does pressure influence the boiling point of water?

Ans.—The higher the pressure, the higher must the temperature of the water be raised in order to cause it to boil.

Ques. 12.—In what way does pressure affect the volume of steam?

Ans.—The higher the pressure, the smaller will be the volume of the steam generated from a given weight of water.

Ques. 13.—In what light should steam be considered relative to work?

Ans.—As an agent through which heat performs the work.

Ques. 14.—What is the most important property of steam?

Ans.—Its expansive force.

Ques. 15.—What law governs this expansion?

Ans.—Boyle's law of expanding gases.

Ques. 16.—Define Boyle's law.

Ans.—The volume of all elastic gases is inversely proportional to their pressure.

Ques. 17.—What is heat?

Ans.—Heat is a form of energy which may be applied to or taken away from bodies.

Ques. 18.—Name the original source of heat, at least for this planet.

Ans.—The sun.

Ques. 19.—How was this heat made available for man's use?

Ans.—By being stored up in oil, wood, and the coal formations millions of years ago, by the rays of the sun.

Ques. 20.—What is the relation of heat to matter?

Ans.—All matter is charged with heat in a greater or less degree, depending upon the nature of the matter.

Ques. 21.—What is the specific heat of any substance?

Ans.—The ratio of the quantity of heat required to raise a given weight of that substance 1 degree in temperature, to the quantity of heat required to raise the same weight of water 1 degree in temperature, the water being at its maximum density, 39.1 degrees.

The following table gives the specific heat of different substances in which engineers are most generally interested:

Table No. 2

Water at 39.1 degrees Fahrenheit	1.000
Ice at 32 degrees Fahrenheit	.504
Steam at 212 degrees Fahrenheit	.480
Mercury	.033
Cast iron	.130

TABLE NO. 2—*Continued*

Wrought iron	.113
Soft steel	.116
Copper	.095
Lead	.031
Coal	.240
Air	.238
Hydrogen	3.404
Oxygen	.218
Nitrogen	.244

Ques. 22.—What is sensible heat?

Ans.—Heat imparted to a body, and warming it. Sensible heat in any substance can be measured in degrees of a thermometer.

Ques. 23.—What is latent heat?

Ans.—Heat given to a body and not warming it; that is, the heat that is not shown by the thermometer.

Ques. 24.—Is the heat lost that thus becomes latent?

Ans.—It is not. On the contrary, it was required to produce the change in the body from the solid to liquid, or from the liquid to the gaseous state. For instance, in the transformation of ice into water, 180 degrees of heat becomes latent, and in changing the water into steam at atmospheric pressure 965.7 degrees of heat become latent.

Ques. 25.—What is the first law of thermo-dynamics?

Ans.—Heat and work are mutually convertible; that is, a certain amount of work will produce a certain amount of heat, and the heat thus produced will, by its disappearance, if rightly applied, produce a fixed amount of mechanical energy.

Ques. 26.—How is heat measured with relation to work?

Ans.—By the thermal unit.

Ques. 27.—What is a thermal unit?

Ans.—It is the quantity of heat required to raise the temperature of one pound of pure water one degree, o. from 39 degrees, its temperature of greatest density, to 40 degrees.

Ques. 28.—What is the mechanical equivalent of heat?

Ans.—The mechanical equivalent of heat is th energy required to raise a weight of 778 pounds one foo high, or a weight of one pound 778 feet high; in othe words, 778 foot pounds. This amount of energy is store in one thermal unit, or heat unit.

Ques. 29.—In how many ways is heat transmitted?

Ans.—In two ways:—First by conduction; second, by radiation.

Ques. 30.—What is conduction of heat?

Ans.—Conduction is the transmission of heat from one body to another in direct contact with it.

Ques. 31.—Are all bodies equally good conductors of heat?

Ans.—No. The best conductors of heat are the metals, silver, copper, tin, steel, lead. The poorest conductors, or nonconductors, as they are termed, are hair, wool, straw, wood, liquids, and "dead" air, that is, air not in circulation.

Ques. 32.—What is radiation of heat?

Ans.—Radiation is the transmission of heat from one

body to another through an intervening space between the bodies.

Ques. 33.—How is the heat in the furnace or fire-box of a boiler transmitted to the water in the boiler?

Ans.—By radiation and conduction through the heating surface of the boiler.

Ques. 34.—What is combustion?

Ans.—Combustion is the chemical union of the carbon and hydrogen of the fuel with the oxygen of the air.

Ques. 35.—What is one of the main factors in the proper combustion of fuels, especially coal?

Ans.—A proper supply of air.

Ques. 36.—What is the principal constituent of coal, oil, and most other fuels?

Ans.—Free or fixed carbon.

Ques. 37.—Are there other combustibles in fuels?

Ans.—Yes; hydrocarbons, a chemical combination of carbon and hydrogen in different ratios.

Ques. 38.—State the composition of air.

Ans.—By volume, 21 parts oxygen and 79 parts nitrogen; by weight, 23 parts oxygen and 77 parts nitrogen.

Ques. 39.—In what proportion do the atoms of carbon and hydrocarbons combine with the atoms of oxygen to form perfect combustion?

Ans.—One atom of carbon combines with two atoms of oxygen, expressed by the chemical symbol CO^2.

Ques. 40.—In the process of combustion, which combustible burns first?

Ans.—When fresh fuel is added to the fire, the hydrocarbons distill in the form of gas, and if the conditions of draught, admission of air, etc., are right, this gas will ignite and burn during its passage through the furnace and combustion chamber; otherwise it passes out of the stack in the form of smoke.

Ques. 41.—What are the common products of combustion?

Ans.—First, carbonic acid, resultant from the chemical union of one atom of carbon with two atoms of oxygen (symbol CO^2); second, water vapor, resultant from the chemical union of two portions of hydrogen, and one portion of oxygen (symbol H^2O); third, inert gases, like nitrogen, also unassociated oxygen, ash, and other products, due to the impurities contained in the coal, or other fuel.

Ques. 42.—In what form does the fixed carbon appear during the process of combustion?

Ans.—After the hydrocarbons have left it, the fixed carbon appears in the form of a glowing mass of coke, uniting with the oxygen to form carbonic acid, and all the heat stored in the carbon is liberated, provided the supply of air is correct; otherwise carbon monoxide (symbol CO) is formed, and only about one-third of the stored heat is liberated, the larger portion of the carbon passing off in the form of soot and smoke.

Ques. 43.—How many thermal units are contained in one pound of carbon?

Ans.—14,500 thermal units.

Ques. 44.—Theoretically, how much air is required for the complete combustion of one pound of coal?

Ans.—By weight, 12 pounds; by volume, 150 cubic feet.

Ques. 45.—Is this law carried out in practice?

Ans.—It is not; a much larger quantity of air (20 to 24 pounds per pound of coal) being supplied in order to insure that all the atoms of carbon may find oxygen.

Ques. 46.—In what two ways is the air supplied to boiler furnaces?

Ans.—First, by natural draught; second, by artificial or forced draught.

Ques. 47.—What causes natural draught?

Ans.—The air in the furnace and uptake becomes heated and consequently much lighter in weight than an equal column of outside air. The heated air is therefore continually rising and passing out of the funnel or smoke-stack, while the outside air rushes into the ash-pit and up through the grates to replace it.

Ques. 48.—How many systems of artificial or forced draught are there?

Ans.—There are two principal systems: First, that in which the air is forced directly into the ash-pits, through conduits leading directly from the fan, or other source of the blast; second, that in which the air is forced directly into the fire-room or stoke-hole, which is made air-tight for this purpose, and from thence the air finds its way into the furnaces on the same principle as when natural draught is employed.

Ques. 49.—Mention the two most important factors in the regulation of combustion.

Ans.—First, the draught; second, the kind and quality of the fuel.

Ques. 50.—What is meant by the expression "rate of combustion?"

Ans.—The rate of combustion means the number of pounds of fuel burned per square foot of grate surface per hour.

Ques. 51.—What are the usual rates of combustion with natural draught?

Ans.—For stationary boilers with shaking grates, from 12 to 18 pounds of coal per hour; for marine boilers, from 15 to 25 pounds.

Ques. 52.—What are the rates of combustion with artificial or forced draught?

Ans.—For stationary boilers, 25 to 35 pounds; for marine boilers, 20 to 50 pounds.

Ques. 53.—How should the air-supply be regulated in order to bring about complete combustion?

Ans.—Complete combustion can be secured only when the air is brought into direct contact, not only with the fuel, but also with the gases as they develop. If the air passing into the furnace above the fuel is first heated, much better results can be attained.

Ques. 54.—Why is it desirable to admit air (heated if possible) above the fire?

Ans.—In order to supply to the hydrocarbons the oxygen necessary to their complete combustion.

Ques. 55.—What will be the result if the supply of oxygen above the fire is not sufficient?

Ans.—A portion of the hydrocarbons will pass off unburned, and of other portions, only the hydrogen is burned, leaving the carbon to pass off as soot or smoke.

Ques. 56.—State another reason why the air should be admitted above the fire.

Ans.—If carbon monoxide (CO) has been formed in the combustion of the fixed carbon, the air above the fire would burn this into carbonic acid, thereby liberating a large additional amount of heat.

Ques. 57.—What is indicated by the formation of much smoke and soot?

Ans.—Incomplete combustion, as smoke and soot are simply unoxydized particles of carbon.

Ques. 58.—Is a high furnace temperature conducive to good combustion?

Ans.—It is; because the hydrocarbons unite with the oxygen much more quickly, and the fixed carbon also is much more completely united with oxygen in a high temperature.

Ques. 59.—Mention a very efficient agency for maintaining a high furnace temperature.

Ans.—Fire-brick arches and bafflers, for the gases to impinge against.

Ques. 60.—Assuming that good combustion is taking place in the boiler furnace, what will be the furnace temperature?

Ans.—From 2,500 to 3,000 degrees Fahrenheit.

Ques. 61.—What are the fuels most commonly used in boiler furnaces?

Ans.—Coal, wood, and oil.

Ques. 62.—What per cent of volatile matter is contained in most of the coals used in the marine service?

Ans.—About 20 per cent.

Ques. 63.—What are the advantages of fuel oil?

Ans.—Greater evaporative power for same weight and bulk, ease of manipulation, perfect control of the combustion to suit requirements of service, and cleanliness.

Ques. 64.—What are the principal objections to the use of oil as fuel for boilers?

Ans.—First, certain dangers involved in storing and using it; second, limited supply.

Ques. 65.—State the difference between the heating value of a pound of bituminous coal and a pound of wood.

Ans.—One pound of coal will evaporate from 8 to 9 pounds of water; one pound of wood will evaporate from 2½ to 3½ pounds of water.

Ques. 66.—What are the two principal kinds of coal used as fuel for boilers?

Ans.—First, anthracite or hard coal; second, bituminous or soft coal.

Ques. 67.—State the composition of hard coal.

Ans.—*Carbon	per cent	91.05
Volatile matter	"	3.45
Moisture	"	1.34
Ash	"	4.16
		100.00

Ques. 68.—State the composition of the best soft coals.

Ans.—†Fixed carbon	per cent	75.02
Volatile matter	"	20.34
Moisture	"	.61
Ash	"	3.47
Sulphur	"	.56
		100.00

*Thurston. †Kent.

Ques. 69.—Does this analysis apply to all bituminous coals?

Ans.—No; some of the poorer kinds run as low as 40 per cent in carbon, 32 per cent in hydrocarbons, and 12 per cent in ash.

Ques. 70.—How do these impurities affect the value of coal as a fuel?

Ans.—The mineral combination of sulphur and iron affects the keeping qualities of some coals. Ashes and mineral substances form clinkers on the grate bars by fusing together, thereby greatly impeding the passage of air through the fire.

TABLE NO. 3

ANALYSIS OF COAL FROM DIFFERENT STATES.

State	Kind of Coal	Moisture	Volatile Matter	Fixed Carbon	Ash	Sulphur
Pennsylvania	Youghiogheny	1.03	36.49	59.05	2.61	0.81
"	Connellsville	1.26	30.10	59.61	8.23	0.78
West Virginia	Quinimont	0.76	18.65	79.26	1.11	0.23
"	Fire Creek	0.61	22.34	75.02	1.47	0.56
E. Kentucky	Peach Orchard	4.60	35.70	53.28	6.42	1.08
"	Pike County	1.80	26.80	67.60	3.80	0.97
Alabama	Cahaba	1.66	33.28	63.04	2.02	0.53
"	Pratt Co.'s	1.47	32.29	59.50	6.73	1.22
Ohio	Hocking Valley	6.59	35.77	49.64	8.00	1.59
"	Muskingum "	3.47	37.83	53.30	5.35	2.24
Indiana	Block	8.50	31.00	57.50	3.00	
"	"	2.50	44.75	51.25	1.50	
W. Kentucky	Nolin River	4.70	33.24	54.94	11.70	2.54
"	Ohio County	3.70	30.70	45.00	3.16	1.24
Illinois	Big Muddy	6.40	30.60	54.60	8.30	1.50
"	Wilmington	15.50	32.80	39.90	11.80	
"	" screenings	14.00	28.00	34.20	23.80	
"	Duquoin	8.90	23.50	60.60	7.00	

Ques. 71.—How is coal measured?

Ans.—Usually by weight in pounds or tons. For storage purposes between 42 and 44 cubic feet per ton of 2,240 pounds are allowed.

Ques. 72.—What is the heating value in thermal units of one pound of bituminous coal?

Ans.—12,000 to 14,500 thermal units, depending upon the quality of the coal.

Ques. 73.—What is the average composition of wood?

Ans.—About 50 per cent of carbon, 40 per cent of oxygen, some hydrogen, and about 1 per cent of ash.

Ques. 74.—What is the average heating value of wood expressed in thermal units?

Ans.—From 6,000 to 8,000 thermal units per pound.

Ques. 75.—What woods are generally used for fuel?

Ans.—Hickory, oak, beech, pines, and firs.

Ques. 76.—What are the principal disadvantages in the use of wood as a fuel for steam boilers?

Ans.—First, a limited supply; second, great bulk in comparison to its heating value.

Ques. 77.—State the composition of fuel oil.

Ans.—Fuel oil contains about 86 per cent of carbon, 13 per cent of hydrogen, and 1 per cent of oxygen.

Ques. 78.—What is the heating value of fuel oil?

Ans.—20,000 to 22,000 thermal units per pound of oil.

Ques. 79.—What are the relative heating values of coal and wood?

Ans.—One pound of coal is equal to 2½ pounds of wood.

Ques. 80.—What is the best all-around fuel, with high heating value, and at reasonable cost?

Ans.—Coal. It is easily obtainable very nearly everywhere; it is safe to handle, and has small bulk in proportion to its heating value.

CHAPTER II

THE BOILER.

Ques. 81.—What are the leading types of boilers in use at the present day in the stationary and marine service?

Ans.—First, fire-tube boilers; second, water-tube boilers.

Ques. 82.—In what respect do they differ?

Ans.—Fire-tube boilers have the hot gases inside the tubes and the water surrounding them, while in water-tube boilers the water is inside the tubes and the hot gases and flame are on the outside.

Ques. 83.—Are boilers classified in any other way?

Ans.—Yes; low-pressure boilers, in which 55 to 60 pounds is the limit, and high-pressure boilers, carrying from 150 to 300 pounds pressure.

Ques. 84.—What are the most common forms of fire-tube boilers?

Ans.—First, the horizontal tubular boiler; second, the vertical tubular boiler; third, the Scotch boiler; fourth, the flue and return tube boiler; fifth, the Western river boiler.

Ques. 85.—Describe the horizontal tubular boiler.

Ans.—It consists of a cylindrical shell, having tubes of from 2 to 4 inches in diameter extending from head to head. There is usually a dome on top, and the boiler is set in brickwork, having the furnace underneath. The heated gases pass first under the boiler, and then return through the tubes to the breeching or uptake leading to the stack.

Ques. 86.—What are the leading features of the vertical tubular boiler?

Ans.—A cylindrical shell, having the fire-box or furnace in its lower end. The bottom ends of the tubes are expanded into the tube-sheet of the fire-box, and the top ends of the tubes are expanded into the top head of the boiler, and conduct the gases directly to the stack.

Ques. 87.—Are the tubes entirely submerged in this class of boilers?

Ans.—Not in all cases. Some forms of vertical boilers have a submerging chamber above the upper tube-sheet. This allows of a steam space above the top ends of the tubes, surrounding the smoke uptake, or smoke flue leading to the stack. The tubes are thus entirely submerged. In the flush-tube boiler the steam and water space is below the upper tube-sheet or head of the boiler, thus leaving the upper portion of the tubes surrounded only by steam.

Ques. 88.—Describe the Scotch boiler.

Ans.—The Scotch boiler may be made either single-ended or double-ended. The shell is cylindrical, with flat heads. The diameters range from 10 to 15 feet, and in some cases even 20 feet, with a length of from 7 to 11 feet. The Scotch boiler is horizontal, and is provided with two or more large corrugated furnace flues, placed near the bottom of the boiler, and extending from the front head to the combustion chamber in the rear.

Ques. 89—What is the diameter of these corrugated flues?

Ans.—From 3½ to 4½ feet, depen ng upon the size of the boiler.

Ques. 90.—How are these furnace flues secured in the boiler?

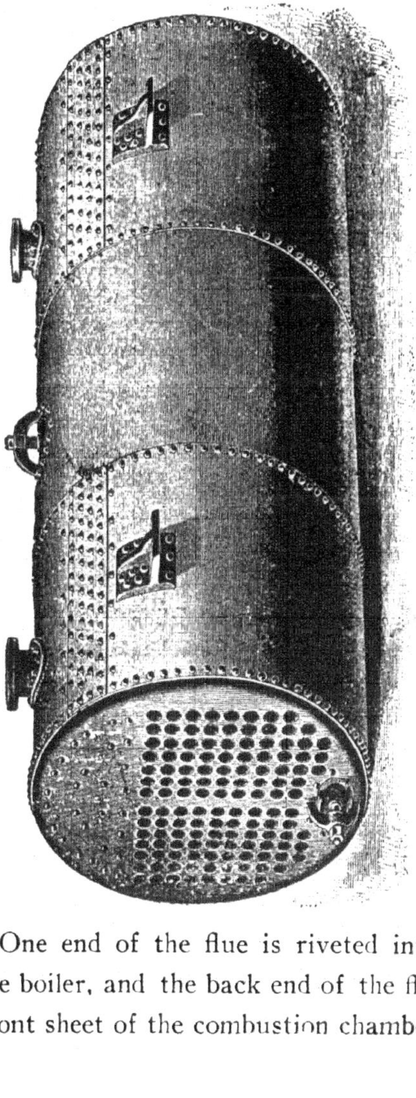

Fig. 1. Horizontal Return Tubular Boiler.

Ans.—One end of the flue is riveted into the front head of the boiler, and the back end of the flue is riveted into the front sheet of the combustion chamber.

Ques. 91.—Describe the combustion chamber of the Scotch boiler.

Ans.—It is a chamber built of steel boiler plate, located at the rear end of the boiler, and entirely surrounded by water. A nest of tubes extends from the front sheet of the combustion chamber, above the corrugated furnace flue, to the front head of the shell.

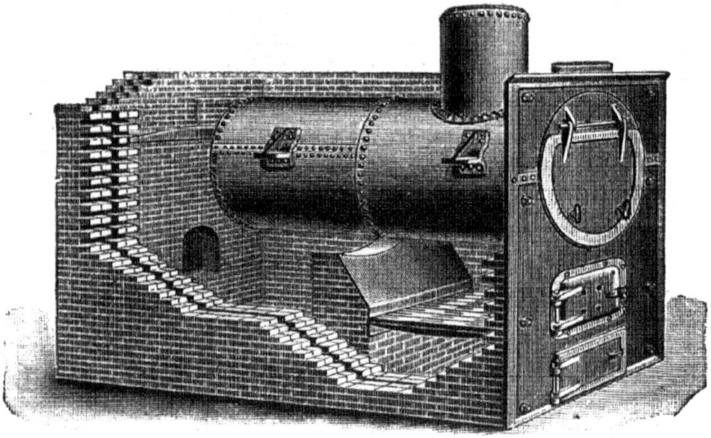

FIG. 2. STANDARD HORIZONTAL BOILER WITH FULL-ARCH FRONT SETTING.

Ques. 92.—Describe the course of the heated gases in the Scotch boiler.

Ans.—The furnaces proper, are placed within the corrugated flues, near the front end. The gases and smoke pass through the flues to the combustion chamber, and from thence return through the small tubes to the smoke-box in front, and from there out through the stack.

Ques. 93.—How are the flat sides of the combustion chamber stayed?

Fig. 3. Vertical Tubular Boiler, with Full-Length Tubes.

Ans.—By stay-bolts connecting with the shell and the back head. The small tubes serve as stays for the front sheet.

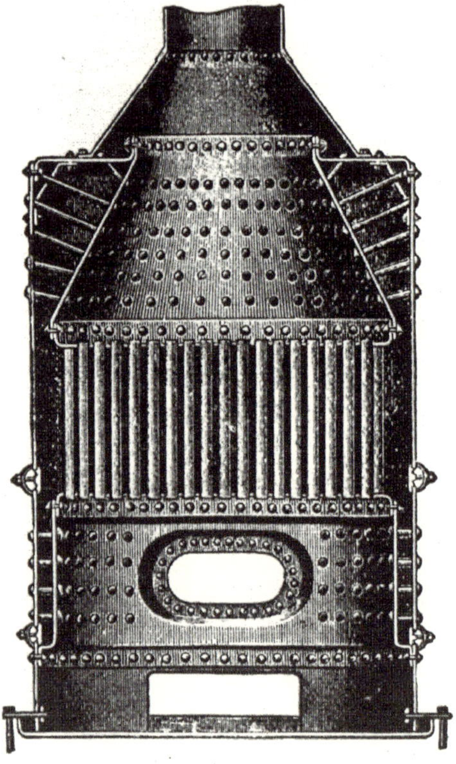

FIG. 4. VERTICAL MARINE BOILER, SHOWING DETAILS OF BRACING.

Ques. 94.—What is meant by double-ended Scotch boilers?

Ans.—Boilers having furnaces at each end. A double-ended Scotch boiler is in fact two single-ended boilers placed back to back.

Ques. 95.—What advantage has the Scotch boiler over other types?

Ans.—A very large amount of heating surface in proportion to its cubic contents.

Ques. 96.—What are the disadvantages connected with the use of the Scotch boiler?

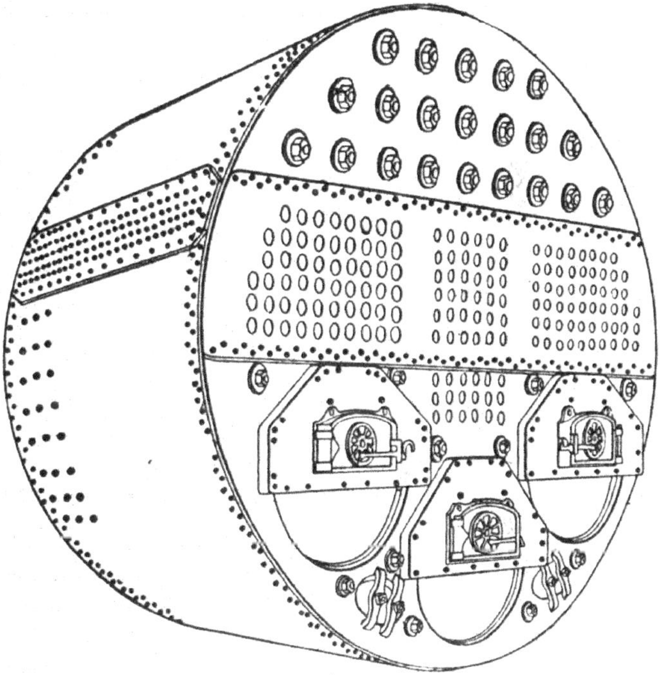

FIG. 5. SINGLE-ENDED SCOTCH MARINE BOILER.

Ans.—First, defective water circulation; second, liability to leaky tubes; third, unequal expansion of the parts, thereby setting up severe strains.

Ques. 97.—Is the Scotch boiler much used?

Ans.—It is in almost universal use in the large ocean-going merchant vessels.

Ques. 98.—What are the distinctive features of the flue and return-tube boiler?

Ans.—This form of boiler is cylindrical in shape in that part of the shell containing the large flues and small return tubes, but resembles a locomotive boiler in that portion containing the fire-box.

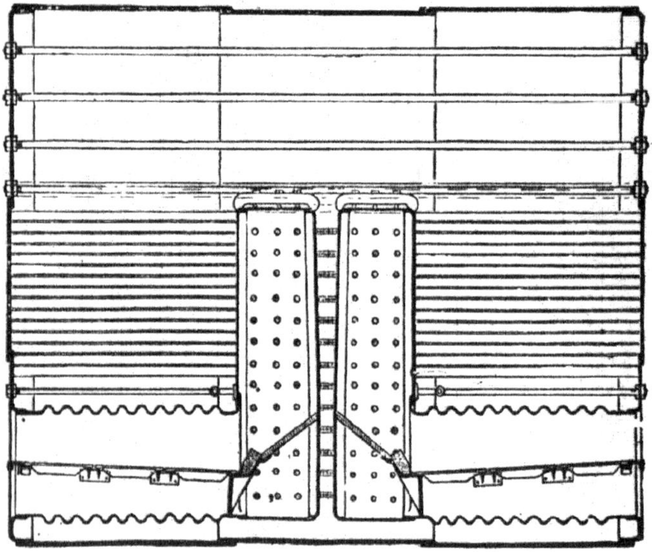

FIG. 6. DOUBLE-ENDED SCOTCH MARINE BOILER, SECTIONAL VIEW.

Ques. 99.—Describe the action of the heat in this boiler.

Ans.—The furnace or fire-box, resembling that of a locomotive, is located in the front end of the boiler. From thence large flues conduct the heated gases to the combustion chamber in the rear, similar to that of a Scotch boiler, and from there the gases return through the small tubes to the uptake.

FIG. 7. RETURN FLUE, CORRUGATED FURNACE BOILER.

Ques. 100.—Describe the Western river boiler.

Ans.—This boiler is usually very long (25 to 30 feet) in proportion to its diameter. It consists of a cylindrical shell having two or more flues of large diameter (12 to 14 inches) extending its entire length. It is set in brick-work in the same manner as the horizontal tubular boiler is, the gases passing underneath the shell to the rear, and thence returning through the large flues to the uptake

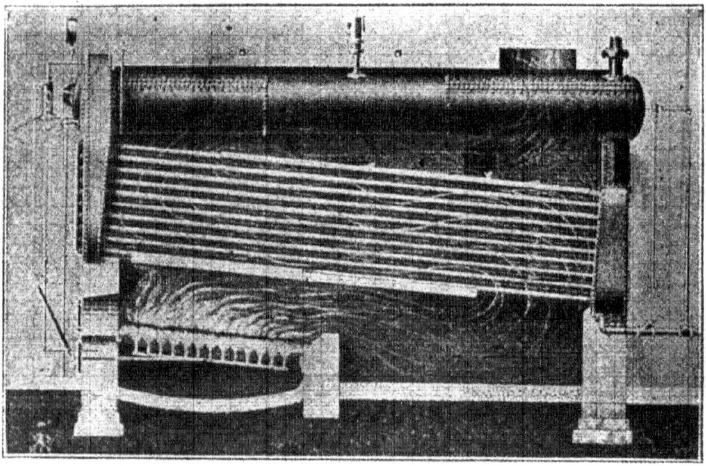

FIG. 8. THE BONUS-FREEMAN WATER-TUBE BOILER.

leading to the stack. It is a very simple boiler, and will withstand high pressures and hard usage.

Ques. 101.—Describe the locomotive boiler.

Ans.—The locomotive boiler consists essentially of a rectangular fire-box and a cylindrical shell. A large number of tubes of small diameter (2 inches) pass through the shell from the fire-box to the smoke-box, a continuation of the barrel at the front end.

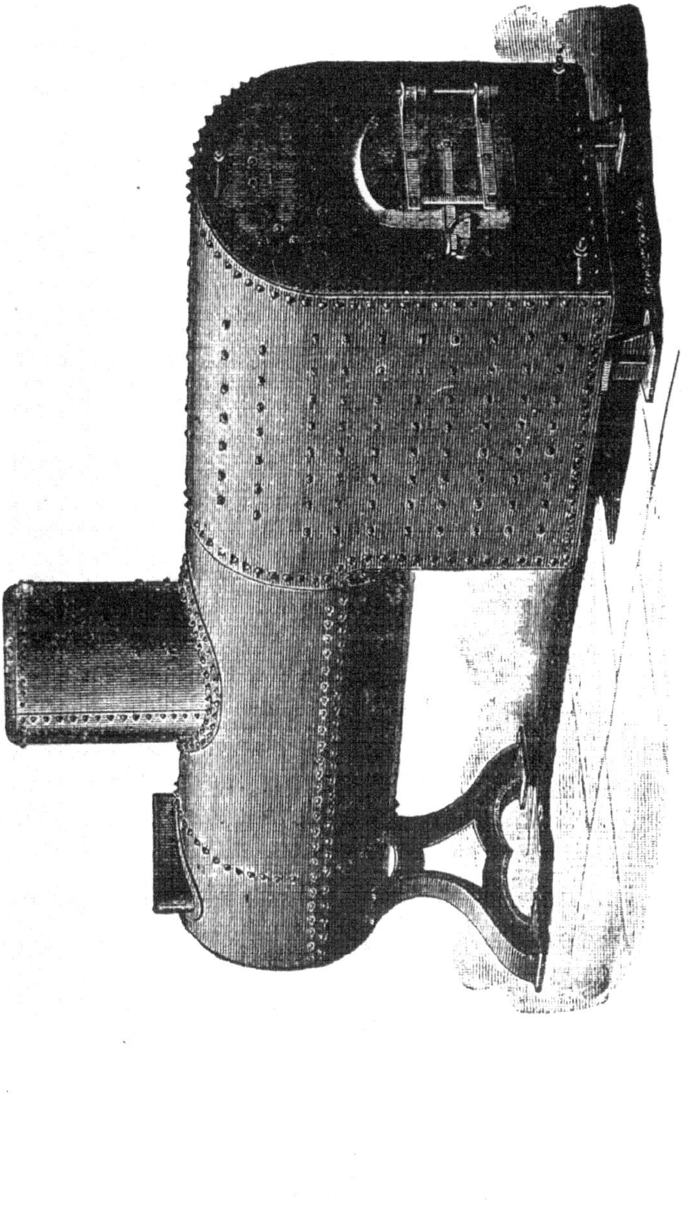

FIG. 9. PORTABLE OR LOCOMOTIVE FIRE-BOX BOILER, WITH WATER FRONT AND OPEN-BOTTOM FIRE-BOX.

Ques. 102.—How is the fire-box joined to the outer shell at the bottom?

Ans.—By a forged ring called the mud-ring, made of wrought iron or steel, through which long rivets pass, uniting the fire-box sheet and the outer sheet.

Ques. 103.—How are the flat sides of the fire-box stayed?

Ans.—By stay-blots screwed through the outer shell, into and through the fire sheet, and having both ends riveted down cold.

Ques. 104.—How is the flat crown-sheet of a locomotive boiler stayed?

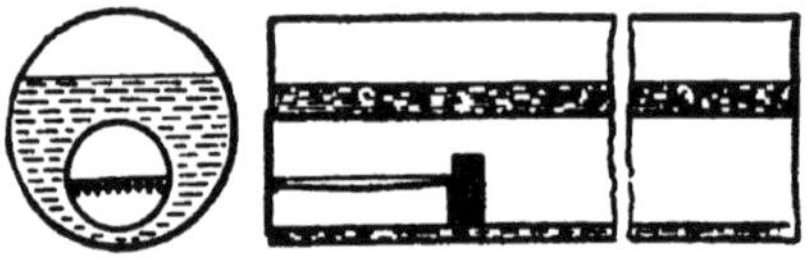

FIG. 10. CORNISH BOILER.

Ans.—By a system of crown-bars, made in the shape of double girders, the ends of which rest upon the side sheets of the fire-box. Crown-bolts pass up through the crown-sheet and crown-bars, and are secured by nuts resting upon saddles on top of the crown-bars. The heads of the bolts support the crown-sheet.

Ques. 105.—Is the locomotive boiler an economical boiler for stationary purposes?

Ans.—It is not.

Ques. 106.—Are there any other forms of cylindrical shell boilers besides those already referred to?

Ans.—Yes; the Cornish boiler, having a large central

flue, in one end of which the furnace is located; the Lancashire boiler, a modification of the Cornish, containing two internal furnace flues, and the Continental boiler.

Ques. 107.—What is meant by Galloway tubes as applied to a boiler?

Ans.—Galloway tubes are conical-shaped water tubes which stand in an inclined position in the large flues of the Lancashire boiler back of the furnaces, and serve to circulate the water from the space below, to the space above the flues. They also act as bafflers to the gases in their passage through the flues, and thus provide increased heating surface.

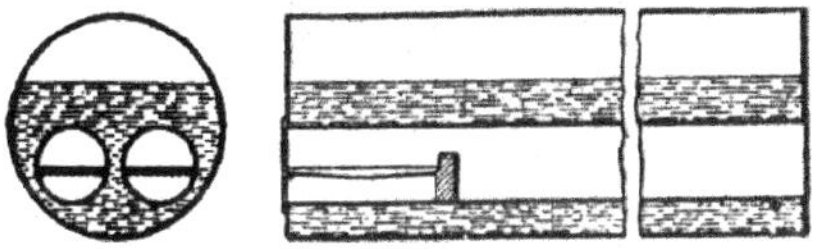

FIG. 11 THE LANCASHIRE BOILER.

FIG. 12. THE GALLOWAY BOILER.

Ques. 108.—Describe the Continental boiler.

Ans.—The Continental boiler is a modification of the Scotch boiler, and is used to a large extent in the marine service. It is provided with a Morison corrugated furnace, and its efficiency as a steam generator has been established by a long series of practical tests.

Ques. 109.—What are the leading characteristics of the Bonson boiler?

Ans.—The Bonson boiler is a combination of the tubular and water-tube types. The water-tube member is in the form of a flat arch, and serves as a roof to the furnace. The cylindrical shell rests upon and is connected with front and rear steel saddles (water-chambers) and the water-tubes are connected with the lower portion of these saddles.

Ques 110.—What route do the gases take in passing from the furnace of the Bonson boiler to the smoke-stack?

FIG. 13. CONTINENTAL BOILER, WITH MORISON CORRUGATED FURNACE, FOR MARINE OR STATIONARY SERVICE.

Ans.—They pass first under the water-tubes, which are lined with a special tile made of fire-clay, the sides of the furnace being also lined with fire-brick. The gases, after passing into the combustion chamber, at the rear, ascend and return through the fire-tubes in the shell, and from thence into the uptake at the front.

Ques. 111.—What are the leading characteristics of water-tube boilers?

Ans.—In water-tube boilers the larger part of the heating surface consists of tubes of moderate size (1 to 4 inches in diameter). There is always some form of separator, drum or reservoir into which the tubes lead. In this drum the steam is separated from the water. In some forms of water-tube boilers this shell or drum is of considerable size.

FIG. 14. THE BONSON BOILER AND SETTING.

Ques. 112.—Is this drum exposed directly or indirectly to the heat?

Ans.—It is generally exposed indirectly, as the upper part is used for steam space.

Ques. 113.—What advantage is there in having a large size steam and water-drum?

Ans.—The advantage of having a good free water surface for the disengagement of the steam. The water occupies about one-third of the lower portion of the drum.

Ques. 114.—Are the upper ends of the tubes in all water-tube boilers entirely filled with water?

Ans.—Not in all cases. In some forms of water-tube boilers the upper ends of the tubes extend above the water level.

Ques. 115.—How are these different forms of water-tube boilers designated?

Ans.—First, as drowned tubes; second, as priming tubes.

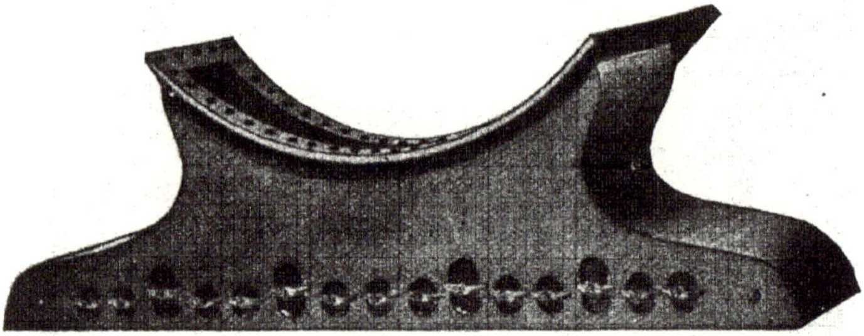

FIG. 15. STEEL SADDLE OF BONSON BOILER.

Ques. 116.—What are some of the advantages of water-tube boilers?

Ans.—They may be made light, powerful and able to withstand high pressures. They are quick steamers, that is, steam may be raised rapidly from cold water; also, the circulation of the water in them is good generally.

Ques. 117.—What are some of the disadvantages attending the use of water-tube boilers?

Ans.—They are difficult to inspect and clean. Also, owing to the large number of joints, leaks are liable to occur.

Ques. 118.—Describe briefly the Babcock & Wilcox water-tube boiler.

Ans.—There is a large horizontal cylindrical shell at the top for the purpose of supplying steam and water-space. The lower half of this shell contains water, and the upper half steam. The tubes are expanded into headers at each end. At the front end these headers are

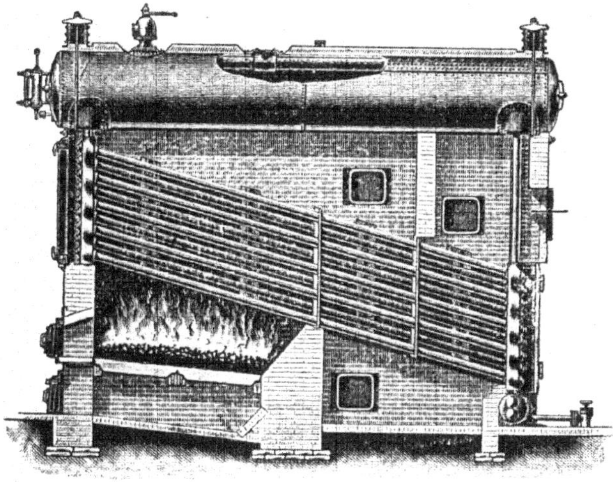

FIG. 16. BABCOCK AND WILCOX BOILER, FOR LAND SERVICE.

brought up near the shell, to which they are connected by a cross connection. The back end headers are connected to a mud-drum at the bottom, and to the shell at the top by slightly inclined tubes. The back headers being lower than the front headers, the tubes are thus inclined from front to back.

Ques. 119.—In what style are the tubes connected to the headers?

Ans.—They are staggered.

Ques. 120.—What is meant by staggered tubes?

Ans.—Staggered tubes are those which are not placed in vertical rows, that is, one directly above the other.

FIG. 17. BABCOCK AND WILCOX "ALERT" TYPE MARINE BOILER. From B. & W. "Book Marine Steam," p 154.

Ques. 121.—What are the facilities for cleaning these tubes?

Ans.—At each end of each tube there are hand-holes provided.

Ques. 122.—Describe the course of the gases for the Babcock & Wilcox boiler.

Ans.—A brick bridge wall at the back end of the furnace, together with special tiles placed among the tubes, compel the gases to first pass up among the tubes until they come in contact with the bottom of the shell for about two-thirds of its length from the front end. At this point a hanging bridge wall and special tiles deflect the gases downward in their course, and they again circulate among the tubes, passing underneath the tiles and up among the tubes again. The products of combustion thus pass over and around the tubes three times on their way to the uptake.

Ques. 123.—What portions of this boiler constitute the heating surface?

Ans.—The tubes, headers, and the lower half of the shell.

Ques. 124.—What course does the water take in its circulation in this boiler?

Ans.—Down from the shell at the rear to the water-tubes, thence forward and upward through the tubes. In its course through the tubes it becomes partially vaporized and of less density. It then passes up into the shell at the front, where the steam is disengaged.

Ques. 125.—Is the Babcock & Wilcox boiler much used in the marine service?

Ans.—Yes, it is used extensively in the British and United States navies, also in merchant steamers.

Ques. 126.—Is the form of this boiler the same for marine as for land service?

Ans.—It is not. The chief features in which it differs from the land boiler are, first, a very much larger grate area; second, the cylindrical shell is set transversely to the direction of the tubes; third, the fire-doors are located at what would be the rear of the land boiler; fourth, the tubes are much shorter, owing to the contracted space allowed on ocean steamers; fifth, the brickwork is surrounded outside by a metal casing.

FIG. 18. THE CALDWELL BOILER.

Ques. 127.—Are there any other forms of water-tube boilers patterned after the Babcock & Wilcox boiler?

Ans.—There are several, prominent among which are the Caldwell and the Root boilers.

Ques. 128.—Describe the Caldwell boiler.

Ans.—It is similar in construction to the Babcock & Wilcox, except that the tubes, instead of being staggered vertically, are placed one directly above the other, with specially shaped fire-brick laid across alternate spaces between the tubes to deflect the gases.

Ques. 129. Describe the Root water-tube boiler.

Ans.—It consists of a nest of 4-inch tubes expanded into headers which are connected at front and back with a set of steam and water-drums about 15 inches in diameter. The tubes are inclined at an angle of about 20 degrees from the horizontal. At the rear end of each overhead water and steam-drum is a connection leading to the

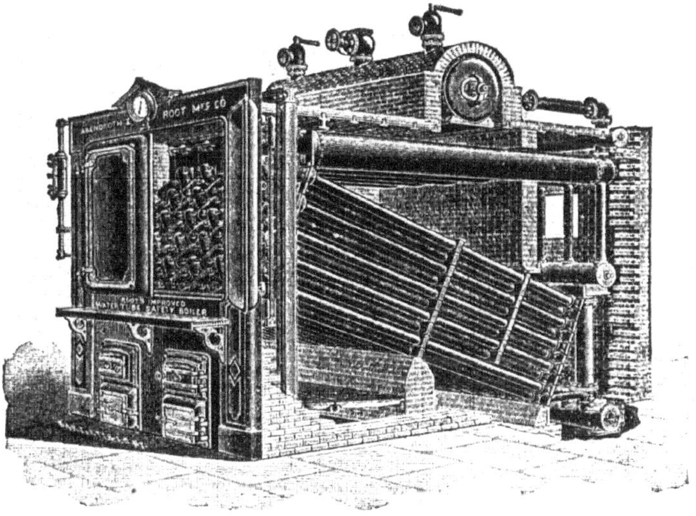

FIG. 19. THE ROOT WATER-TUBE BOILER.

steam-collecting header above, placed transversely to the direction of the other drums, and from this header two connecting pipes lead to a large steam-drum located at about the center of the boiler, and above all.

Ques. 130.—How does the water circulate in the Root boiler?

Ans.—It descends through vertical connecting pipes from the feed-drum at the rear to the mud-drum beneath.

From thence it passes into the back and lower ends of the tubes, and on up through the tubes, and into the overhead drums, into the upper halves of which the steam is disengaged.

Ques. 131.—Describe the Cahall water-tube boiler.

Ans.—The Cahall boiler is vertical, having a nest of

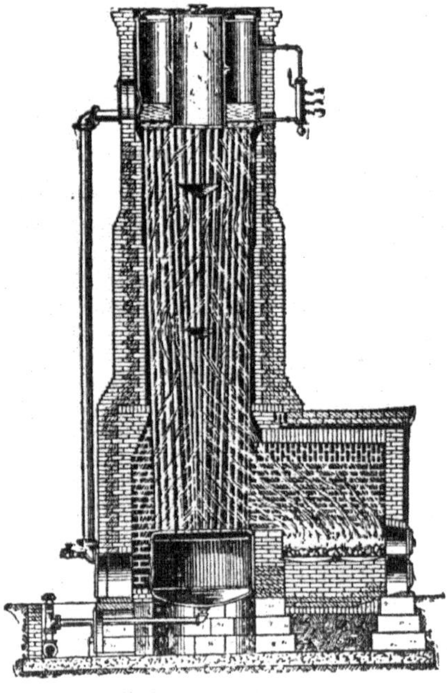

FIG. 20. THE CAHALL VERTICAL BOILER.

water-tubes standing nearly vertical. These tubes are connected with a shallow water-drum at the bottom, and a larger and deeper water and steam-drum at the top. The furnace is located alongside of the mud-drum, and the gases traverse among the tubes in a circuitous manner owing to bafflers placed among the tubes.

Ques. 132.—How do the gases escape to the stack in this boiler?

FIG. 21. WICKES VERTICAL WATER-TUBE BOILER.

Ans.—Extending through the center of the annular drum at the top is a flue through which the products of combustion find their way to the uptake.

Ques. 133.—Of what form is the Wickes boiler?

Ans.—The Wickes boiler consists of upper and lower vertical drums connected by vertical tubes. The furnace is external.

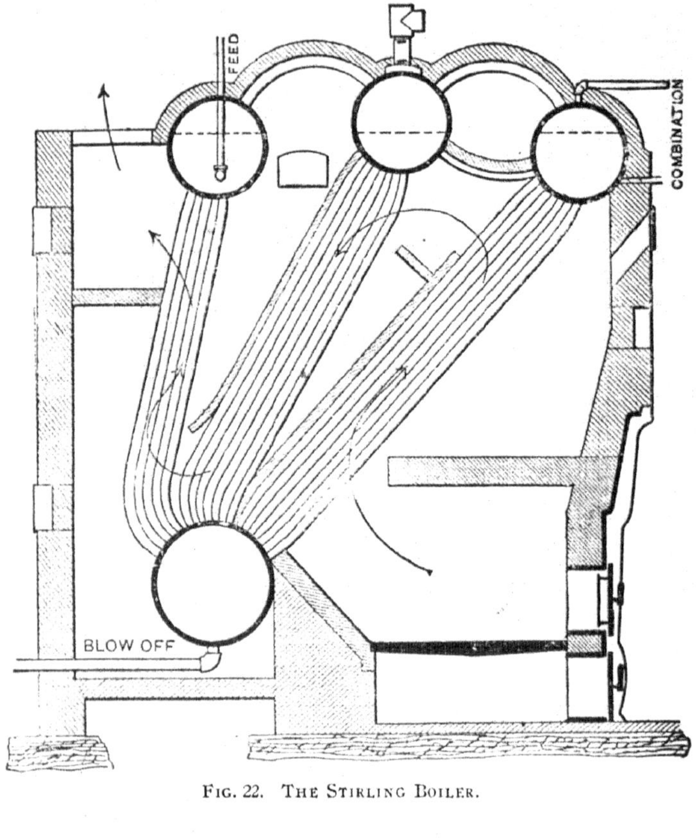

FIG. 22. THE STIRLING BOILER.

Ques. 134.—What course do the gases take in their passage to the stack, in the Wickes boiler?

Ans.—A thin partition wall of fire-brick is built between two adjoining middle rows of tubes. This wall causes the gases first to ascend to the top, and then downwards to the chimney flue at the bottom and opposite to the furnace.

Ques. 135.—Describe the Stirling boiler.

Ans.—In the Stirling water-tube boiler there are three horizontal steam and water-drums at the top, and

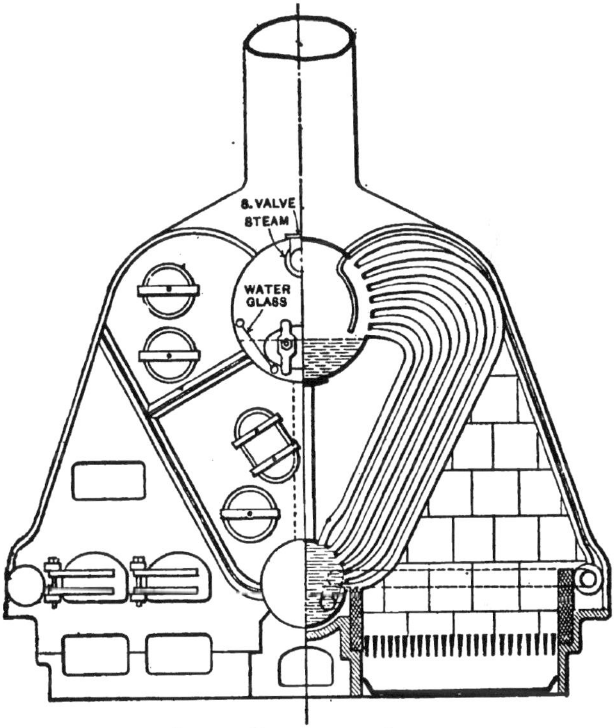

FIG. 23. THORNYCROFT BOILER.

one water-drum at the bottom. These drums are connected by three divisions of inclined and curved tubes.

Ques. 136.—How are the products of combustion led from the furnace to the uptake, in the Stirling boiler?

Ans.—Bafflers of fire-brick are placed back of the two first divisions of tubes. The first baffler causes the gases to ascend to the top of the first division of tubes; the second baffler deflects the gases downwards, around and among the tubes of the second division. The draught is then upwards again, surrounding the tubes composing the third division, thence to the stack.

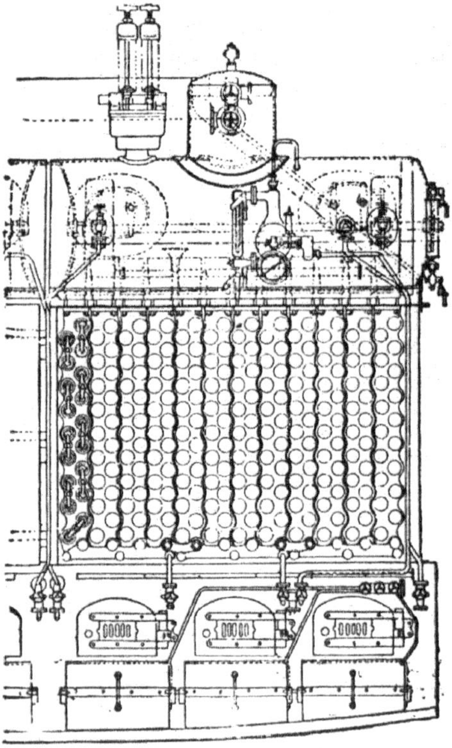

FIG. 24. THE NICLAUSSE BOILER.

Ques. 137.—Describe the Thornycroft boiler.

Ans.—The Thornycroft boiler is adapted for use on torpedo boats and high-speed yachts. A large horizontal steam-drum at the top is connected to a water-drum at the bottom by two groups of curved tubes of small diameter. The grates are located on each side of the water-drum. There are also two smaller drums at the bottom, one on each side, connected to the middle drum by small pipes.

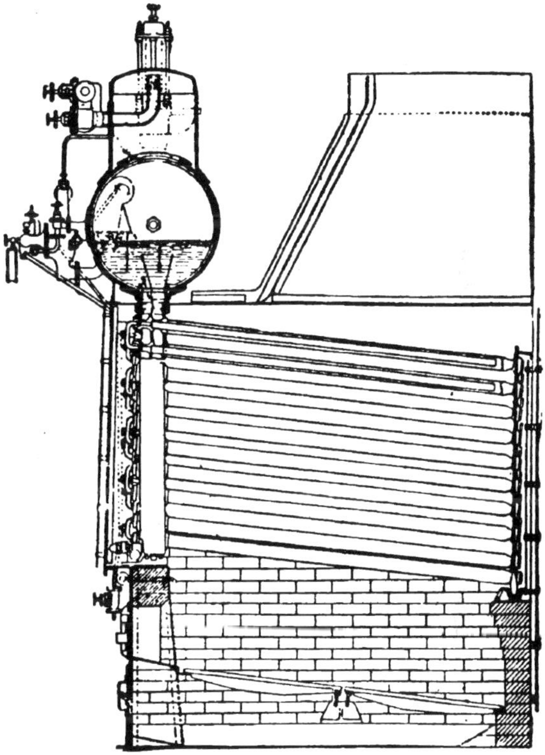

FIG. 25. THE NICLAUSSE BOILER—SIDE VIEW.

Ques. 138.—How does the water circulate in this boiler?

Ans.—Down from the top drum to the middle lower drum through special return water-tubes of large diameter, and from thence through the smaller tubes to the side drums. From there the water passes up through the curved tubes to the upper portion of the top drum, where the steam is disengaged.

Ques. 139.—Describe the Niclausse boiler.

Ans.—The Niclausse boiler is made up of a series of slightly inclined tubes. These tubes are double, that is, one inside the other, and they are connected to the front header in such a manner that the colder water flows down the inside tubes and returns to the front between the two tubes when heated by the action of the fire and hot gases on the larger outside tubes. Each vertical row of tubes is connected at the front end to a separate header, the headers being placed side by side, and all leading into a top drum or steam-collector.

Ques. 140.—How is the entering feed-water at the front kept separate from the hot ascending currents of water?

Ans.—By a diaphragm in the top drum that keeps the cooler water separate from the hot water and steam.

Ques. 141.—How are the tubes connected to the headers in the Niclausse boiler?

Ans.—By coned surfaces on the ends of the tubes bearing on similar coned surfaces in the headers, and kept in contact by outside dogs and nuts. These joints appear to cause no trouble by leakage.

Ques. 142.—Is the Niclausse boiler much used?

Ans.—It is used to some extent in the British navy, and also in several large United States war-ships.

Ques. 143.—Of what type is the Normand boiler?

Ans.—The Normand boiler is a marine water-tube

boiler of the Thornycroft type. The two outer rows of tubes are formed into a wall of tubes, and in the vicinity of the furnace the tubes are arched upwards in order to form a combustion chamber. Back of the furnace the

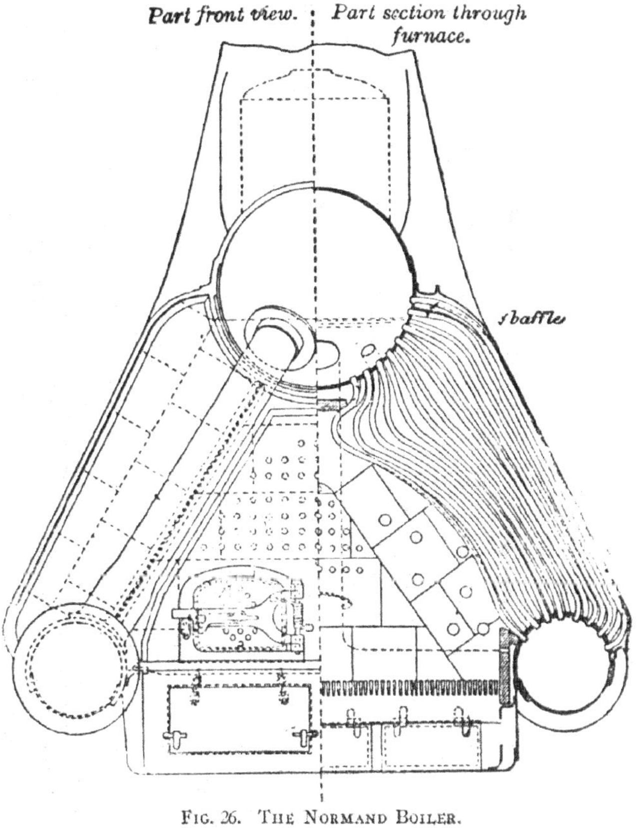

FIG. 26. THE NORMAND BOILER.

curvature is not so great, although all of the tubes are curved more or less, to permit of expansion when heated.

Ques. 144.—What course do the gases take in this boiler?

Ans.—The gases proceed from the fire among the tubes, and traverse the length of the boiler to the rear end, where they pass below a brick deflecting plate to the space surrounding those tubes that are less curved.

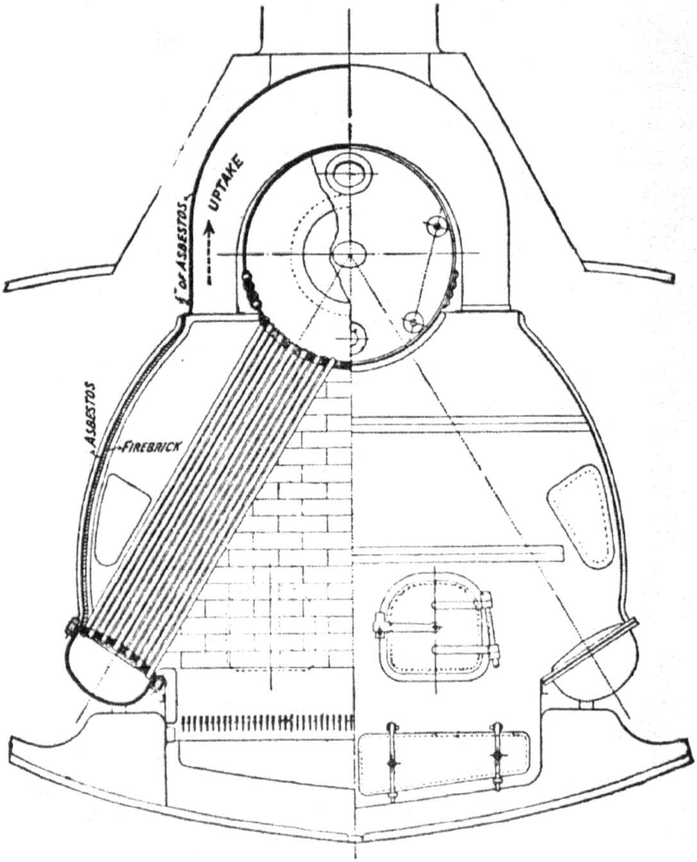

FIG. 27. THE YARROW BOILER.

Ques. 145.—What other peculiar feature characterizes the Normand boiler?

Ans.—Provision is made for the admission of air above the fire.

Ques. 146.—How is this accomplished?

Ans.—By means of a small air casing at the front and back, and a series of small holes one inch in diameter leading through the brickwork to the space above the fire.

Ques. 147.—For what kind of service is the Normand boiler mainly adapted?

Ans.—For torpedo-boat destroyers.

Ques. 148.—What is the distinguishing feature of the Yarrow boiler, among boilers having water-tubes of small diameter?

Ans.—The Yarrow boiler has straight tubes. It also has at the bottom on each side a small water-chamber or mud-drum with nearly flat tube-plates, into which the tubes are expanded. The tubes run in an inclined direction from these water-drums to the steam and water-drum at the top.

Ques. 149. In what manner does the water circulate in the Yarrow boiler?

Ans.—Those tubes which receive the most heat conduct the water from the lower drums to the upper drum, into which the steam is delivered. Other tubes which are cooler carry the water from the upper drum to the lower drums.

Ques. 150.—Describe the Mosher boiler.

Ans.—The Mosher boiler has two upper steam-drums and two lower and smaller water-drums, the water-drums being directly underneath the steam-drums. These drums are connected by curved generator pipes of small diameter, the pipes entering the steam-drums above the water-line.

Ques. 151.—How does the water find its way from the upper to the lower drums?

Ans.—By means of two external downtake pipes 4 inches in diameter. The boiler is cased in, the casing being lined with fire-brick.

Ques. 152.—For what class of service is the Mosher boiler mainly adapted?

Ans.—Torpedo boats and high-speed yachts.

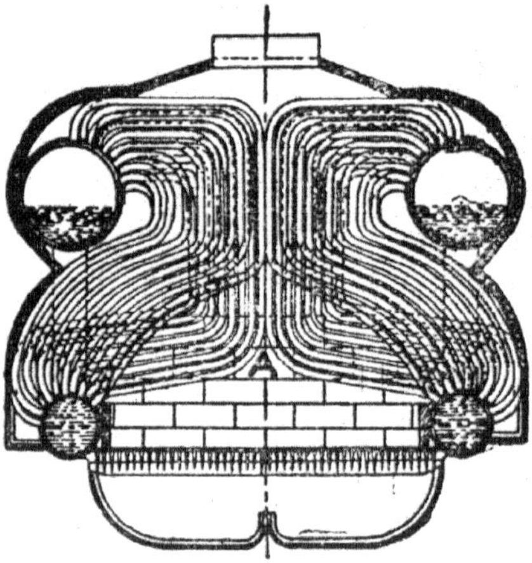

FIG. 28. THE MOSHER BOILER.

Ques. 153.—Describe the construction of the Almy boiler.

Ans.—It is made principally of short lengths of pipe screwed into return bends and into twin unions. At the bottom there is a larger pipe or header that surrounds the two sides and back of the grates, and there is a similar structure at the top, the two headers being connected by the smaller pipes.

Ques. 154.—How is the steam separated from the water in the Almy boiler?

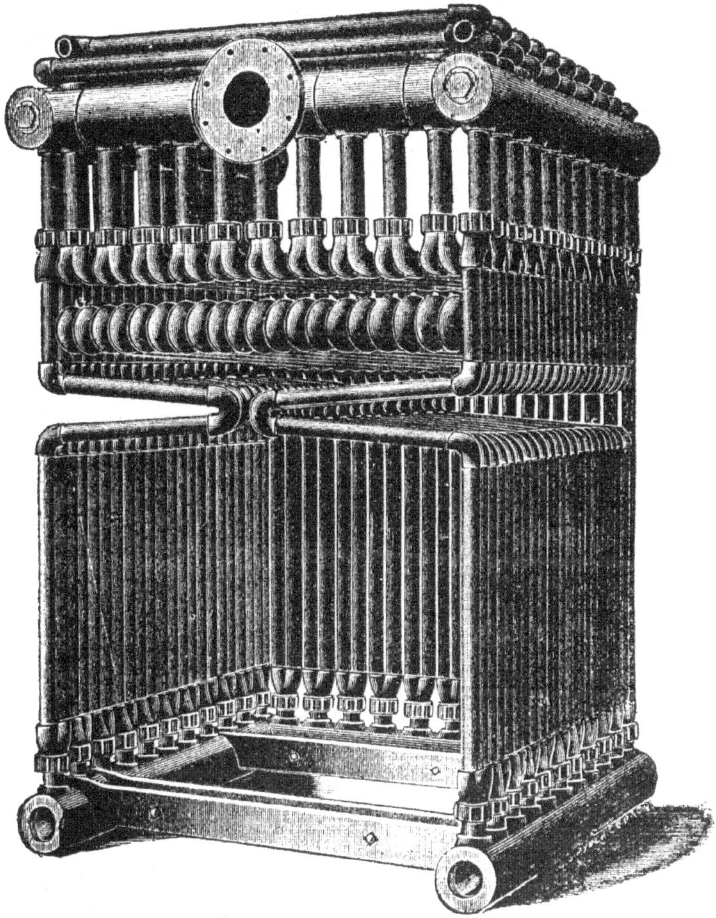

FIG. 29. THE ALMY BOILER.

Ans.—The steam and water are together discharged from the upper header into a separator in front of the boiler, and from this separator the steam is drawn, while the separated water and the feed-water pass down through circulating pipes to the lower header.

Ques. 155.—What other peculiar feature attaches to this boiler?

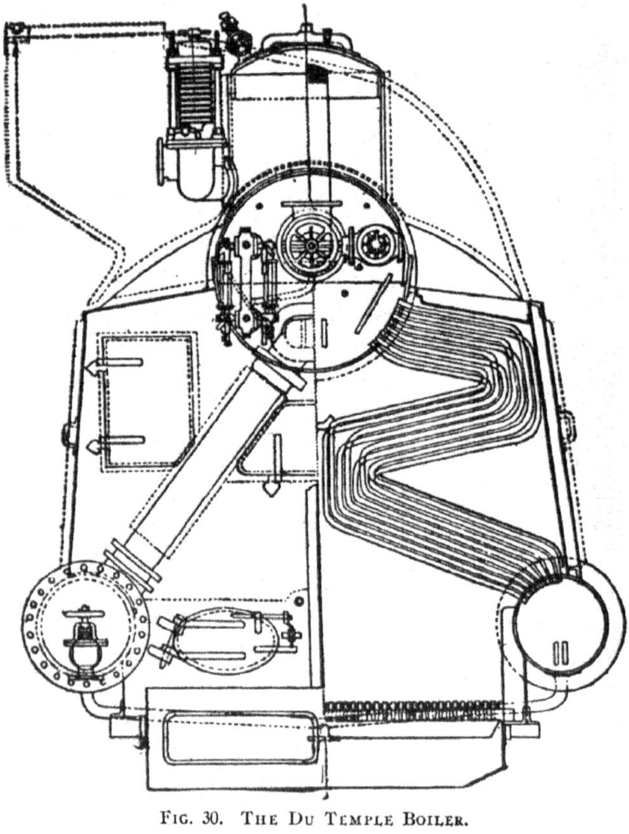

FIG. 30. THE DU TEMPLE BOILER.

Ans.—It is provided with a coil feed-water heater above the main boiler.

Ques. 156.—Describe in general terms the Du Temple boiler.

Ans.—It is of the same general character as the Thornycroft type, except that the generating tubes discharge into the steam-drum below the water-line.

Ques. 157.—How are these tubes connected to the drums?

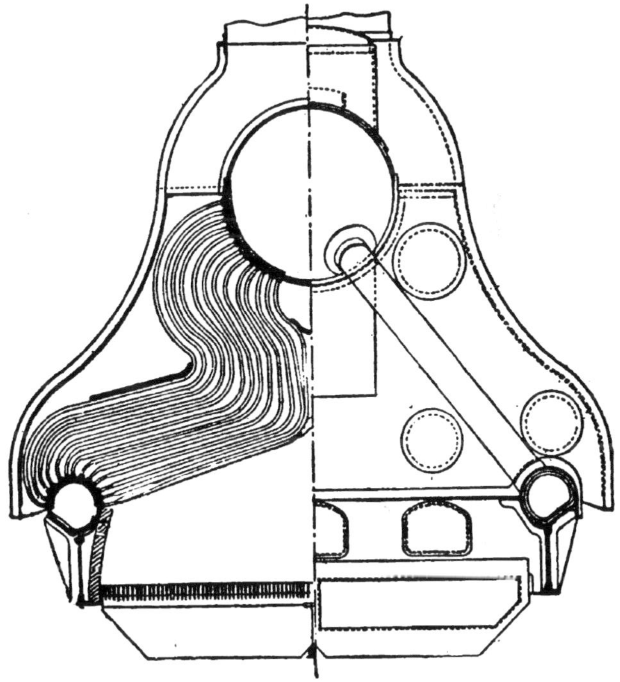

FIG. 31. REED'S BOILER.

Ans.—By cones and nuts.

Ques. 158.—Is the Du Temple boiler used to any great extent?

Ans.—Yes; it is used extensively in the French navy, especially on vessels of the torpedo-boat type.

Ques. 159.—Describe Reed's boiler.

Ans.—This boiler resembles the Du Temple boiler. It has the usual top collector drum, and two lower drums with curved generating pipes connecting them.

Ques. 160.—How are the tubes attached to the drums?

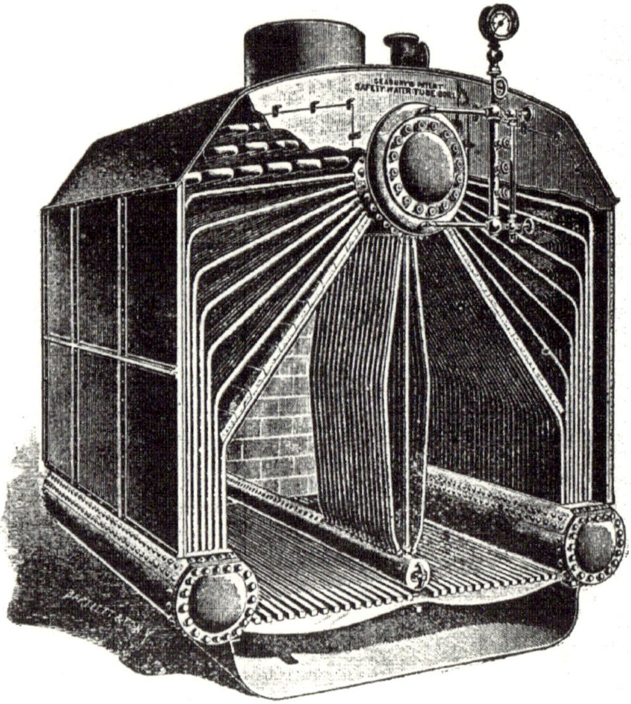

Fig. 32. The Seabury Boiler.

Ans.—By screwed connections at each end, with nuts inside the chambers.

Ques. 161.—How are the gases caused to traverse the heating surface in this boiler?

Ans.—By means of diaphragms fitted to the tubes.

Ques. 162.—What class of service is this boiler largely used in?

Ans.—British torpedo-boat destroyers, and also on third-class cruisers.

Ques. 163.—Describe the Seabury boiler.

Ans.—The Seabury boiler has three lower water-drums, the middle drum being smaller than the two outside drums. These drums are connected to one large steam and water-drum above by curved pipes of small diameter and the furnace is divided into two sections by the central nest of pipes. Above the boiler tubes and inside the casing there is a coil feed-water heater.

Ques. 164.—Describe the latest type of Belleville boiler?

Ans.—The Belleville boiler is a water-tube boiler, and is of extensive use on large ships. It is made up of two distinct series of straight tubes, larger in diameter than those of the curved type. These tubes are placed nearly horizontal, each alternate horizontal row being slightly inclined in the opposite direction to the row above it. The generator proper has a water-chamber below and a steam-drum or chamber on top, and the zigzagged tubes are connected to these respective chambers.

Ques. 165.—What kind of a furnace has this boiler?

Ans.—A rectangular brickwork furnace inclosed in a steel casing, and the generating tubes are placed directly over the grates, the bottom row of tubes being about two feet above the grates. Baffle plates are secured at intervals among the tubes for the purpose of causing the hot gases to traverse the whole of the heating surface.

Ques. 166.—How is circulation of the water secured in the Belleville boiler?

Ans.—By means of external return water-pipes, one on each side connecting the ends of the top drum with the lower water-chamber, the cooler water thus passing

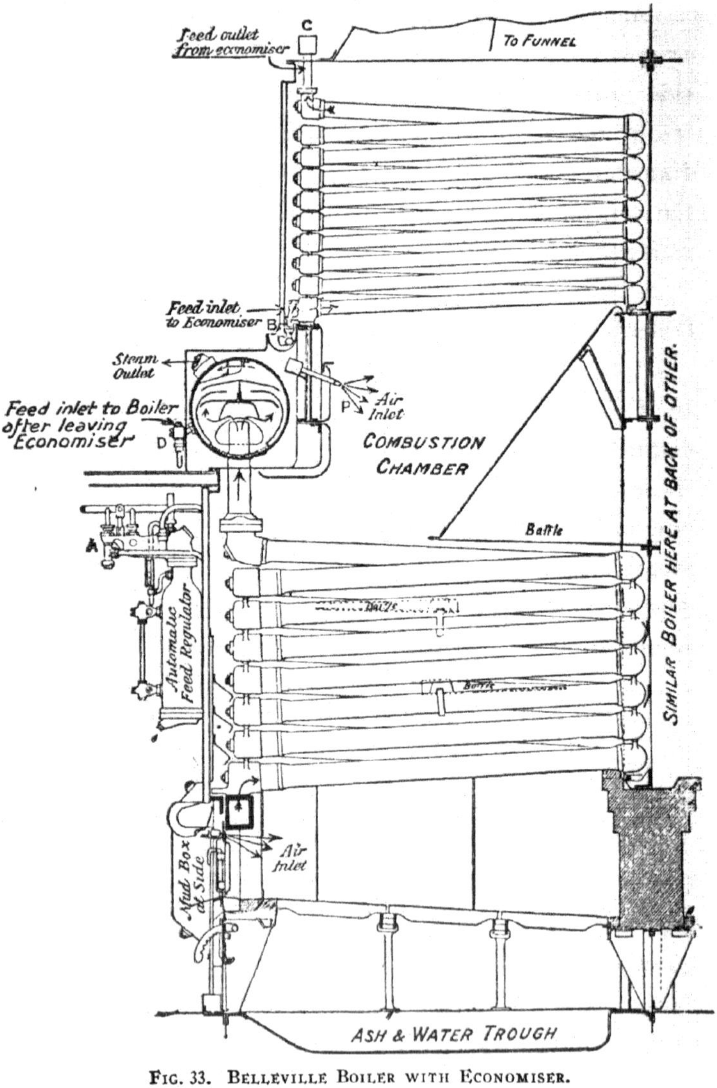

Fig. 33. Belleville Boiler with Economiser.

down through these pipes into the lower drum, and from thence the heated water passes up through the generating tubes, discharging into the top drum, where the steam is disengaged.

Ques. 167.—What are the usual dimensions of the generating tubes?

Ans.—Four and one-half inches in diameter and seven feet six inches in length. The ends are connected by being screwed into malleable cast-iron boxes.

Ques. 168.—How is the economizer or feed-water heater attached to this boiler?

Ans.—It is placed directly above the generator, a space called the combustion chamber being left between the two series of tubes. The tubes of the economizer are smaller, being 2¾ inches in diameter. The general form of the economizer resembles that of the generator.

Ques. 169.—What is the course of the feed-water in this boiler?

Ans.—It enters the bottom of the economizer and is forced upwards to and fro through the zigzagged tubes to the top, and from thence it falls to the bottom of the hot water collector at the top, and then flows to the return pipes, through which it passes to the generator.

Ques. 170.—Mention another peculiar feature of this boiler.

Ans.—An automatic feed-regulating device worked by a float in a chamber acting upon the feed-valve.

Ques. 171.—Is the Belleville boiler an economical boiler?

Ans.—It is; an actual evaporation of from 9.3 pounds

to 9.9 pounds of water per pound of coal having been obtained under test, with the feed-water at a temperature of 68 degrees.

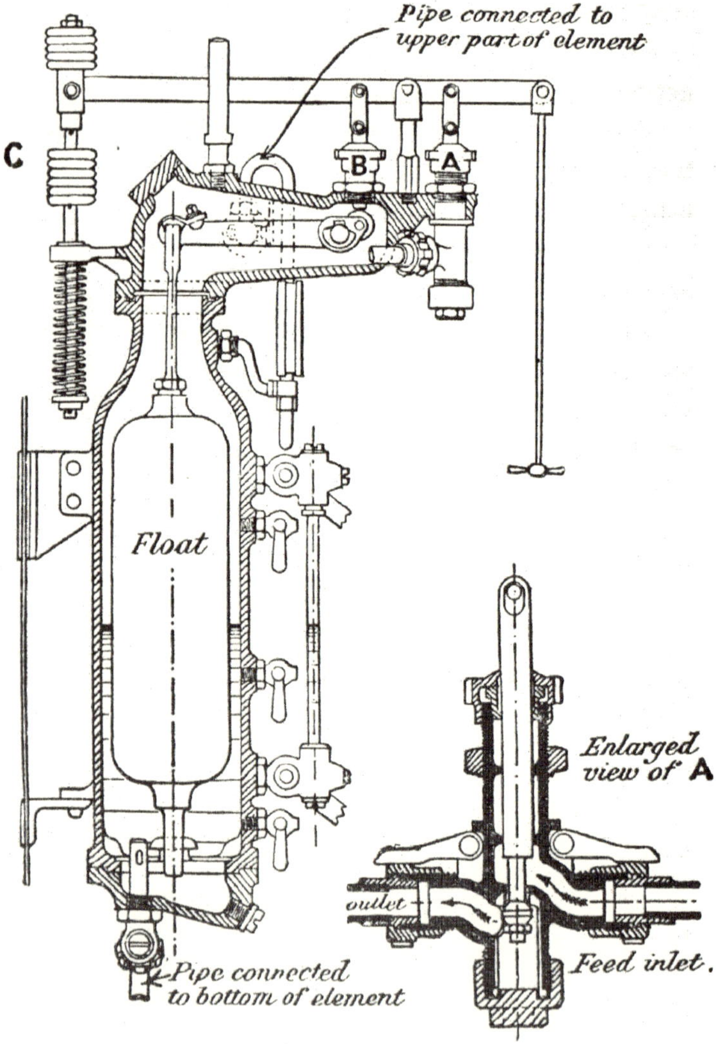

FIG. 34. AUTOMATIC FEED REGULATOR FOR BELLEVILLE BOILER.

CHAPTER III

BOILER CONSTRUCTION

Ques. 172.—What is the best material to use in the construction of the shell of the boiler?

Ans.—Open-hearth steel, having a tensile strength of from 55,000 pounds to 60,000 pounds per square inch.

Ques. 173.—What is meant by the expression tensile strength (T. S.)?

Ans.—The expression 60,000 pounds tensile strength means that it would require a pull of 60,000 pounds in

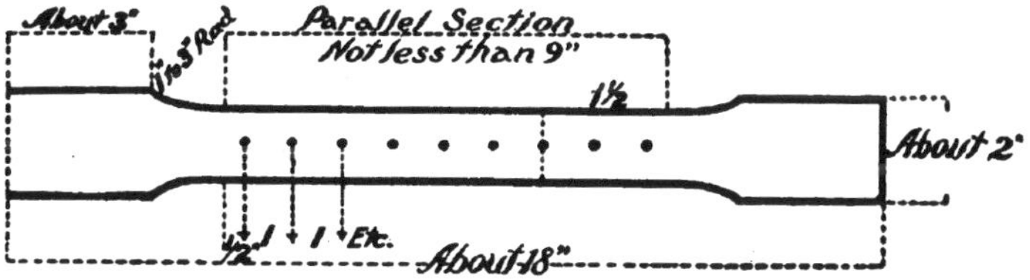

FIG. 35. TEST PIECE.

the direction of its length to break a bar of the material 1 inch square, or 2 inches wide by ½ inch thick, or 2.67 inches wide by ⅜ inch thick.

Ques. 174.—How are steel sheets for boiler construction tested?

Ans.—A small piece, called a test piece, is cut from each sheet and placed in a testing machine.

Ques. 175.—What is the working test for steel boiler sheets?

Ans.—A piece from each sheet is heated to a dark

cherry red, plunged into water at 60° temperature, and bent double cold under the hammer, such piece to show no flaw or crack after doubling.

Ques. 176.—Of what material should the tubes of fire-tube boilers be made?

Ans.—A good quality of homogeneous iron.

Ques. 177.—What is the working test for boiler tubes?

Ans.—They should show no flaw when expanded into the flue-sheet and beaded.

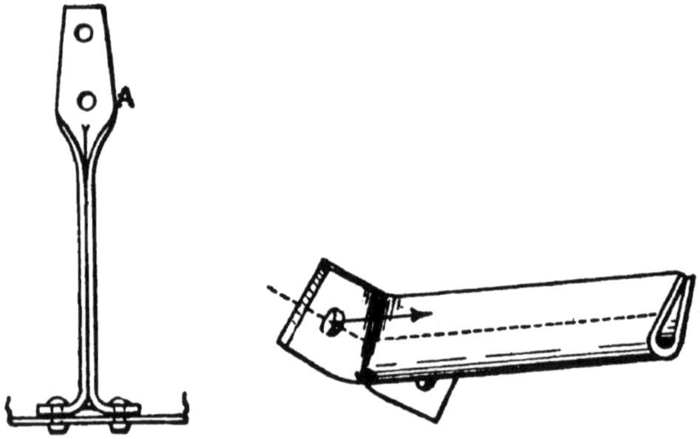

Fig. 36. Crow Foot Braces.

Ques. 178.—What should the specifications be regarding rivets?

Ans.—All rivet material should be of good charcoal iron, or mild steel, tough and soft. Test, a good rivet should bend double cold, without showing fracture.

Ques. 179.—Of what material are the tubes of water-tube boilers usually made?

Ans.—Of good charcoal iron or mild steel specially prepared for the purpose, and lap welded, or drawn.

Ques. 180.—What is the test for tubes from 3½ to 4 inches in diameter and No. 10 wire gauge?

Ans.—A piece 1½ inches in length is cut from one end of a tube, and this piece must stand hammering down cold vertically without showing a crack or split, when down solid.

Ques. 181.—Of what material should stay-bolts be made?

Ans.—Of iron or mild steel, especially manufactured for the purpose.

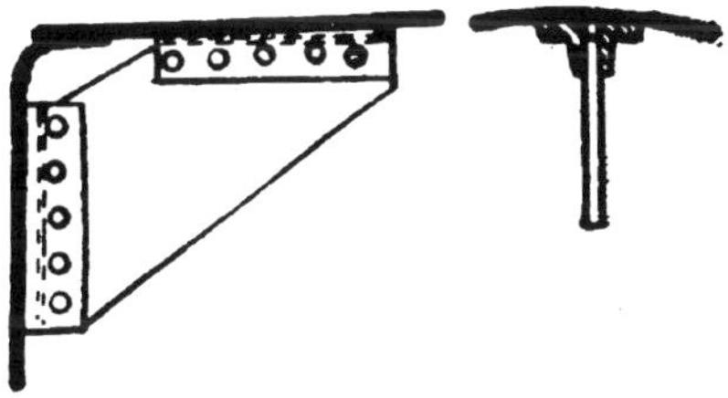

FIG. 37. GUSSET STAYS.

Ques. 182.—What should be the tensile strength of stay-bolt material?

Ans.—For iron, not less than 46,000 pounds; for steel, not less than 55,000 pounds.

Ques. 183.—What kind of material are braces and stays made of?

Ans.—The material for braces and stays should be of the same quality as the best stay-bolt stock.

Ques. 184.—What is the object sought in staying the flat surfaces of a boiler internally?

Ans.—The object is to strengthen those surfaces sufficiently to enable them to withstand the maximum internal working pressure to which they will be subjected.

Ques. 185.—Does the cylindrical portion of a boiler need bracing?

Ans.—It does not, for the reason that the internal pressure tends to keep it cylindrical.

Ques. 186.—What is the maximum direct pull per square inch of section that may be allowed on braces and stay-rods?

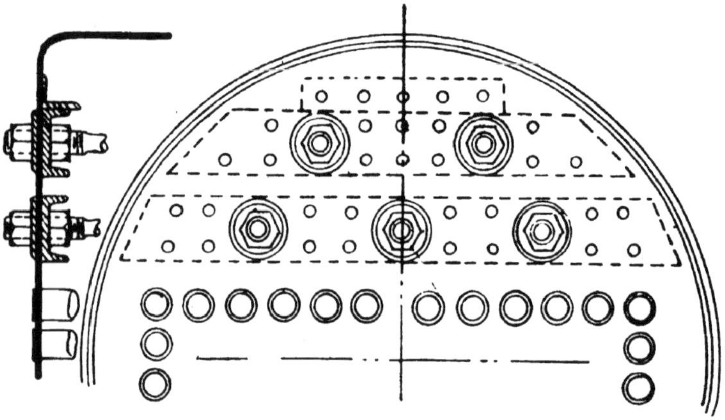

FIG. 38. THROUGH STAY RODS.

Ans.—For iron, 6,500 pounds; for steel, 8,000 pounds; and this point should be kept in view when spacing the braces.

Ques. 187.—What is meant by spacing braces?

Ans.—The distance from center to center that the stays are from each other at the point of their connection to the stayed surface.

Ques. 188.—Give an example.

Ans.—The stays in a certain boiler are spaced 8 inches apart, center to center, therefore each stay supports

8 x 8=64 square inches. Assuming the working pressure to be 100 pounds per square inch, the sectional area of each stay should be 1 square inch.

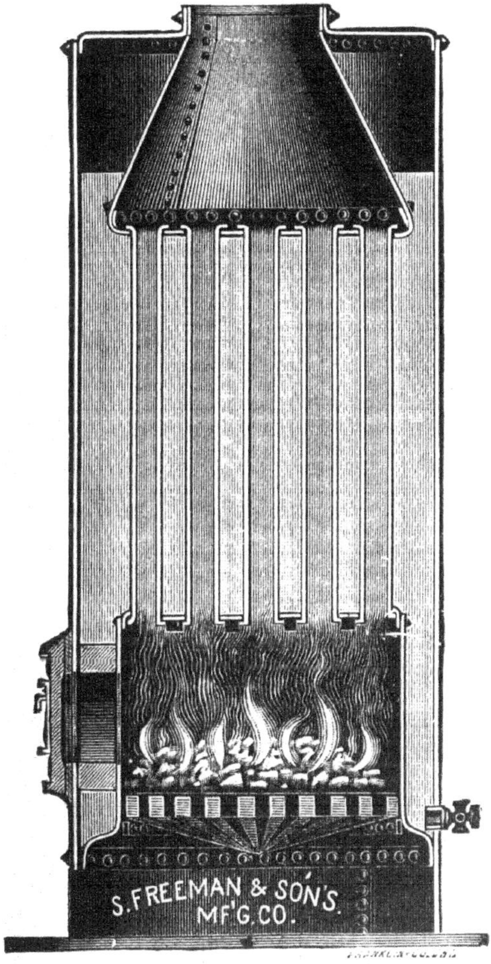

FIG. 39. VERTICAL TUBULAR BOILER, WITH SUBMERGED TUBES.

Ques 189. —Suppose the working pressure is 250 pounds per square inch and the stays are spaced 6 inches center to center, what should be the sectional area of each stay?

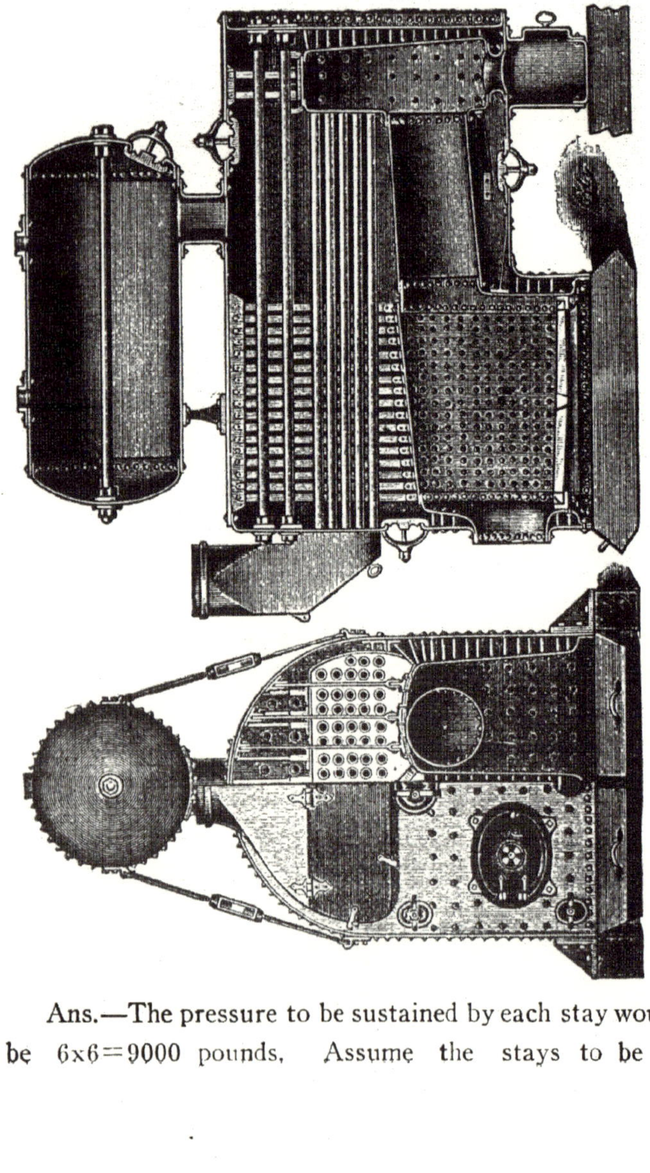

FIG. 40. DOUBLE FURNACE RETURN FLUE MARINE BOILER.

Ans.—The pressure to be sustained by each stay would be $6 \times 6 = 9000$ pounds, Assume the stays to be of steel and unwelded, and allowing a direct pull of 7,200 pounds per square inch, the sectional area of each stay should be $\frac{9000}{7200}=1.25$ square inches; or, if the stays are 1.5 inches smallest diameter, and a direct pull of 8.000 pounds per square inch of section is allowed, they may be spaced 7 inches, center to center.

Ques. 190.—Of what forms are boiler stays usually made?

Ans.—For low-pressure boilers, crow-foot stays; for high-pressure boilers, through stay-rods and gusset-stays.

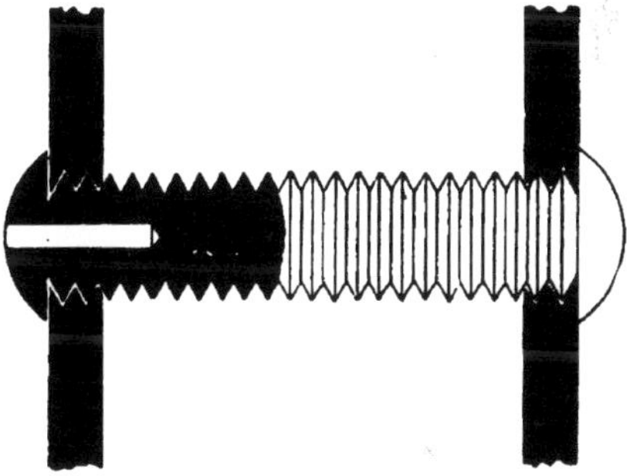

Fig. 41 Common Stay-Bolt.

Ques. 191.—Where are stay-bolts used?

Ans.—In fire-box boilers, and all boilers of the locomotive type, to tie the fire-box to the external shell.

Ques. 192.—How are stay-bolts applied?

Ans.—A continuous thread is cut on the stay-bolt rod, the same thread being also tapped in the holes in the external plate, and the inside sheet. The steel stay-bolt is then screwed through the plates and allowed to project far enough at each end to permit of its being riveted down cold,

Ques. 193.—What is the principal cause of the breaking of stay-bolts?

Ans.—The unequal expansion of the sheets into which they are screwed.

Ques. 194.—Why are stay-bolts sometimes drilled partly through their length?

Ans.—In order that, if the bolt breaks, the steam or water may blow out through the small hole and give warning of the break.

Ques. 195.—Describe the Tate flexible stay-bolt.

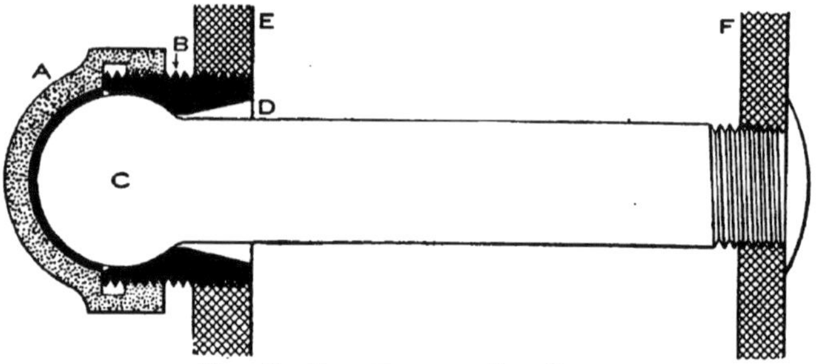

FIG. 42. TATE FLEXIBLE STAY BOLT.

Ans.—The outer head is ball shaped, and is inclosed within a socket formed by a sleeve that screws into the outer sheet and a cap that screws onto the sleeve. The other end of the bolt is screwed into and through the fire-sheet a sufficient distance to allow of riveting.

Ques. 196.—What is meant by the efficiency of a riveted joint?

Ans.—It is the per cent. of strength of the solid plate that is retained in the joint.

Ques. 197.—What is the efficiency of a properly proportioned double riveted butt-joint?

Ans.—From 71 to 75 per cent.

Ques. 198.—What is the efficiency of a properly proportioned triple riveted butt-joint with inside and outside welts or butt-straps?

Ans.—From 85 to 88 per cent.

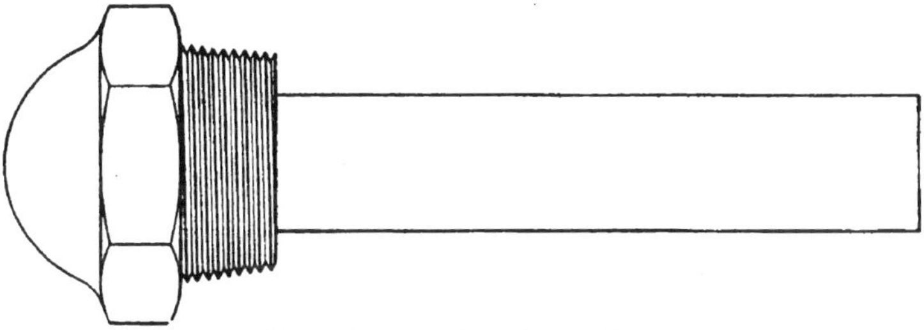

FIG. 43. TATE FLEXIBLE STAY-BOLT, UNTHREADED.

Ques. 199.—Where is the weakest portion of the triple riveted butt-joint?

Ans.—At the outer row of rivets.

TABLE 4

TABLE OF DIAMETERS OF RIVETS*

Thickness of Plate	Diameter of Rivet	Thickness of Plate	Diameter of Rivet
1/4 inch	1/2 inch	9/16 inch	7/8 inch
5/16 "	9/16 "	5/8 "	15/16 "
3/8 "	11/16 "	3/4 "	1 1/16 "
7/16 "	3/4 "	7/8 "	1 1/8 "
1/2 "	13/16 "	1 "	1 1/4 "

*Machine design—W C. Unwin.

Ques. 200.—What percentage of efficiency may be retained in a properly designed quadruple riveted butt-joint having both inside and outside butt-straps?

Ans.—94 per cent.

Ques. 201.—Where is the weakest portion of such a joint?

Ans.—At the outer row of rivets.

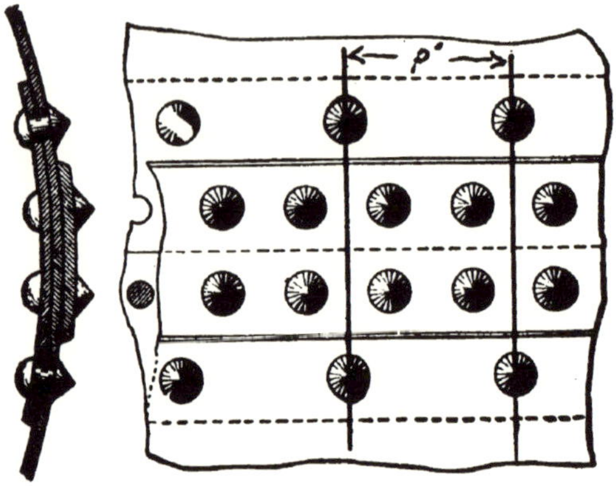

FIG. 44. DOUBLE RIVETED BUTT-JOINT.

Ques. 202.—How may boiler heads be constructed which will not require to be stayed?

Ans.—By being dished, or "bumped up."

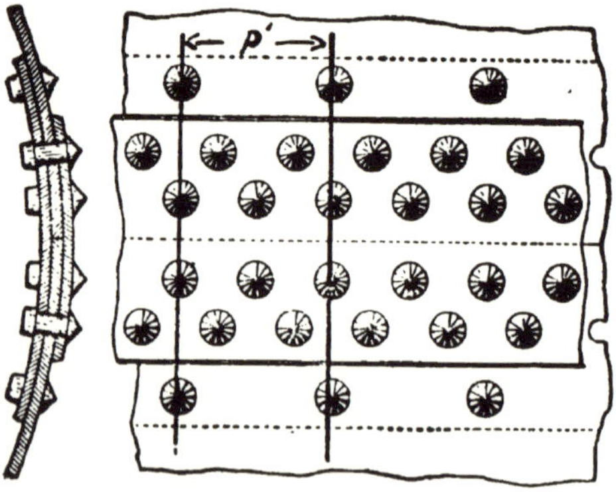

FIG. 45. TRIPLE RIVETED BUTT-JOINT.

Ques. 203.—What is the depth of dish, as adopted by steel-plate manufacturers?

Ans.—One eighth of the diameter of the head, when flanged.

Ques. 204.—What should be the thickness of the head as compared to the thickness of the shell?

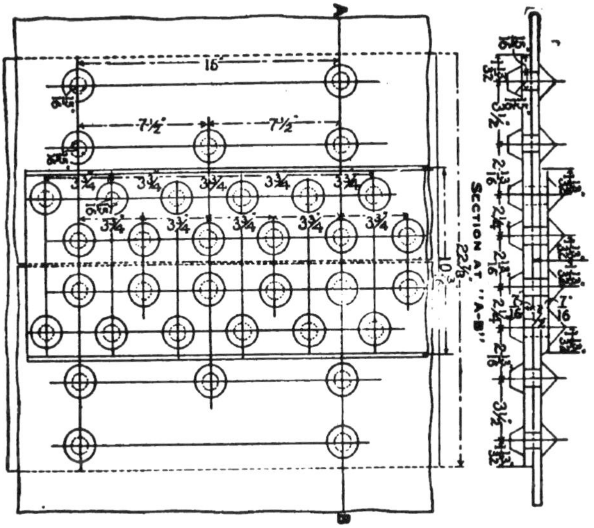

FIG. 46. QUADRUPLE RIVETED BUTT-JOINT.

Lloyd's rules, condensed, are as follows:

LLOYD'S RULES—THICKNESS OF PLATE AND DIAMETER OF RIVETS

Thickness of Plate	Diameter of Rivets	Thickness of Plate	Diameter of Rivets
3/8 inch	5/8 inch	3/4 inch	7/8 inch
7/16 "	5/8 "	13/16 "	7/8 "
1/2 "	3/4 "	7/8 "	1 "
9/16 "	3/4 "	15/16 "	1 "
5/8 "	3/4 "	1 "	1 "
11/16 "	7/8 "		

Ans.—The heads should be as thick, or slightly thicker, than the shell plate.

Ques. 205.—What method other than riveting may be, and sometimes is employed in the formation of boiler seams?

Ans.—Boiler seams may be welded if the material from which the plates are rolled is of the best, and great care and skill are exercised.

Ques. 206.—Mention two of the advantages possessed by welded seams over riveted seams?

TABLE 5

PROPORTIONS OF TRIPLE-RIVETED BUTT JOINTS WITH INSIDE AND OUTSIDE WELT

Thickness of Plate Inches	Diameter of Rivet Inches	Pitch of Rivet Inches	Pitch of Outer Rows Inches	Efficiency Per Cent
3/8	13/16	3.25	6.5	84
7/16	13/16	3.25	6.5	85
1/2	13/16	3.25	6.5	83
9/16	7/8	3.50	7.0	84
5/8	1	3.50	7.0	86
3/4	1 1/16	3.50	7.0	85
7/8	1 1/8	3.75	7.5	86
1	1 1/4	3.87	7.7	84

Ans.—First, a good welded joint approaches more nearly to the full strength of the material than can possibly be attained by rivets, no matter how correctly designed the riveted joint may be; second, the welded joint, having a smooth surface inside the boiler, is much less liable to collect scale and sediment than is the riveted joint.

Ques. 207.—Why should the longitudinal or side seams of a boiler be stronger than the girth or round-about seams?

Ans.—Because the force tending to rupture the boiler along the line of the longitudinal seams is proportional to the diameter divided by two, while the stress tending to pull it apart endwise is only one-half that, or proportional to the diameter divided by four.

Ques. 208.—What is the formula for ascertaining the bursting pressure of a boiler?

Ans.—$\frac{TS \times T \times E}{R} = B$, in which

TS = Tensile strength
T = Thickness of sheet
E = Efficiency of joint
R = Radius (one-half the diameter)
B = Bursting pressure

Ques. 209.—How is the safe working pressure of a boiler ascertained?

Ans.—First calculate the bursting pressure, then divide this by the factor of safety, which usually is five, although in some instances a safety factor of eight is used.

Ques. 210.—In addition to the regular bracing and staying, how are the heads of return tubular and Scotch marine boilers greatly reënforced?

Ans.—By the tubes, which are expanded into the heads and beaded down on the ends.

Ques. 211.—Are the tubes always expanded into the tube-sheets?

Ans.—They are in fire-tube boilers. In some forms of water-tube boilers the tubes are screwed into the headers or chambers.

Ques. 212.—What type of furnace is largely used in internally fired boilers?

Ans.—The Morison corrugated furnace.

Ques. 213.—Mention three advantages gained by the use of corrugated furnaces.

Ans.—First, the corrugations (if properly made) add great rigidity and strength to resist the crushing strain to which the furnaces are subjected; second, there is more heating surface in a corrugated than in a smooth surface; third, the alternate expansion and contraction of the

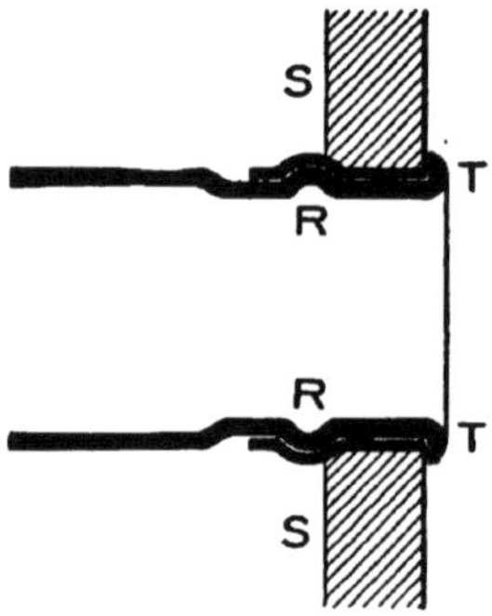

FIG. 47. SECTION OF TUBE EXPANDED INTO SHEET.

corrugated surface tends to loosen any scale that may form on the surface inside the boiler.

Ques. 214.—In regard to riveted seams, which is the better method, to drill or to punch the rivet-holes?

Ans.—The rivet-holes should be drilled. In good boiler work this method is now always followed.

Ques. 215.—What other important point should be kept in view in joining the plates of a boiler?

Ans.—To get the joint tight without caulking, or at least with as small an amount of caulking as possible.

Ques. 216.—Mention some of the injurious effects of excessive caulking.

Ans.—First, it is one of the most fruitful causes of grooving along the edges of the seams; second, it tends to raise the edge of the plate that is caulked, thereby causing looseness at the joint.

Ques. 217.—What other very important point should be secured in the construction of the boiler?

Ans.—The rivet-holes in the plates should come *fair* before the rivet is put in.

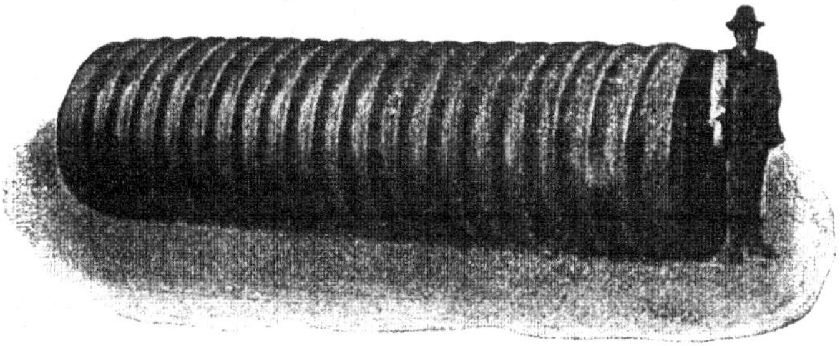

FIG. 48. MORISON CORRUGATED FURNACE.

Ques. 218.—If the rivet-holes do not come *fair* what should be done with them?

Ans.—They should be made exactly true by the use of a rimer.

Ques. 219.—What should not be done with the rivet-holes in case they do not come *fair?*

Ans.—They should not be drifted. A drift-pin is often the primary cause of starting a crack in a sheet.

Ques. 220.—What can be said generally concerning the construction of a boiler, especially one intended for high pressures?

Ans.—Only the best material should be used, and great care and skill should be exercised in all the detail of assembling it.

By reference to Chapter I, Part 2, of Swingle's "Twentieth Century Hand Book for Engineers and Electricians," the student will be enabled to obtain much more detailed information concerning boiler construction, the strength of riveted joints, bracing and staying, strength of material, etc., as all of these important features are dwelt upon at length and fully discussed.

—

CHAPTER IV

BOILER SETTINGS AND APPURTENANCES.

Ques. 221.—What kind of a setting is required for internally fired boilers?

Ans.—First, a good solid foundation, second, the boiler should be covered with non-conducting, non-combustible material of some sort, to prevent radiation of heat, and the whole should be encased in a sheet-metal jacket.

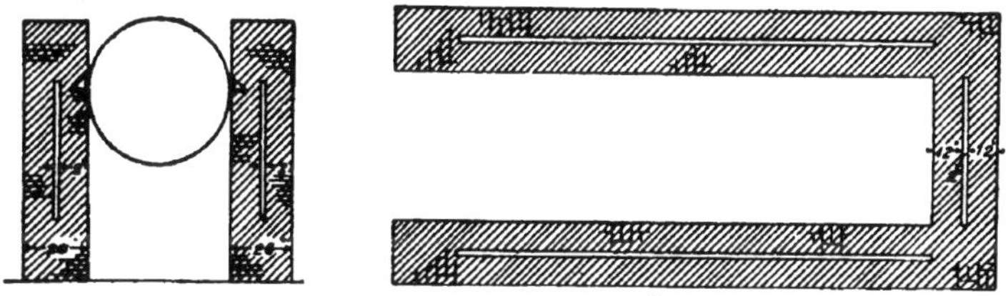

FIG. 49. PLAN AND ELEVATION OF BOILER SETTING, SHOWING AIR SPACES.

Ques. 222.—What kind of a setting is required for horizontal tubular and water-tube boilers?

Ans.—Brick walls with an inner lining of fire brick. When the boiler is supported by lugs resting upon the walls, a heavy iron plate should be imbedded in the brickwork, for each lug to rest upon. The walls should also be tied together, both endwise and transversly, by iron rods not less than 1¼ inch in diameter, extending clear through in both directions, the bottom rods to be laid in place as the walls are being built. These rods are to have a thread and nut on each end, and are secured to heavy cast or wrought iron bars called buck stays, placed vertically against the outside of the walls.

Ques. 223.—How may boiler walls be greatly protected from the injurious action of the heat?

Ans.—By leaving an air-space of 2 inches between the fire-brick lining and the outer wall, beginning at the level of the grate bars and extending as high as the center of the boiler. Above this height the walls should be solid.

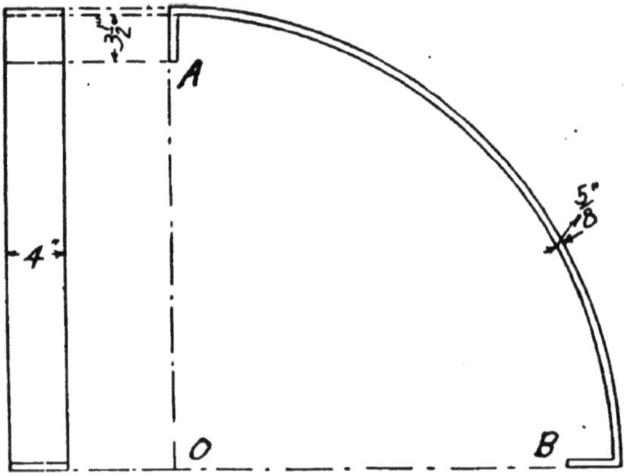

Fig. 50. Clamp for Back Arch.

Ques. 224.—What is the duty of bridge-walls and bafflers?

Ans.—To present a hot surface for the unconsumed gases to impinge against, and also to divert the gases towards the heating surface of the boiler.

Ques. 225.—How may a good and durable back arch for a horizontal tubular boiler be constructed?

Ans.—Take flat bars of iron ⅝ inch thick by 4 inches in width, cut them to the proper length, bend them to the shape of an arch, and turn 4 inches of each end back at right angles. The clamp thus formed is to be filled with a course of side arch fire-brick, and will form a complete and self-sustaining arch 9 inches wide and with sufficient spring to cover the distance between the back wall and the back head of the boiler above the tubes. Enough of these arches should be made so that when laid side by side they will cover the distance from one side wall to the other, across the rear end of the boiler.

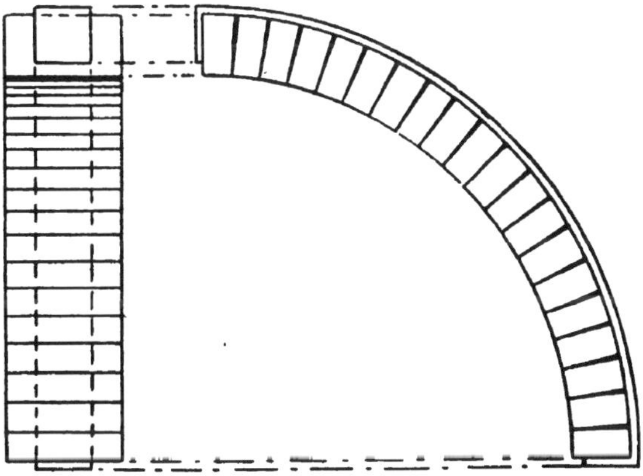

FIG. 51. BACK ARCH COMPLETE.

Ques. 226.—What advantages do this form of back arch possess over the ordinary flat cover?

Ans.—First, it can come and go with the expansion and contraction of the boiler; second, it always maintains a practically air-tight cover at this important point; third, in case of needed repairs to the back end of the boiler the sections may be easily removed, one at a time, and when the repairs are completed they may be reset with very small expense.

Ques. 227.—Give an easy rule for ascertaining the dimensions of the grates.

Ans.—For a horizontal tubular, the length of the grates should equal the diameter of the boiler. The width depends upon the construction of the furnace. If the fire-brick lining is built perpendicular, the width of grate will also equal the diameter of the boiler, but if

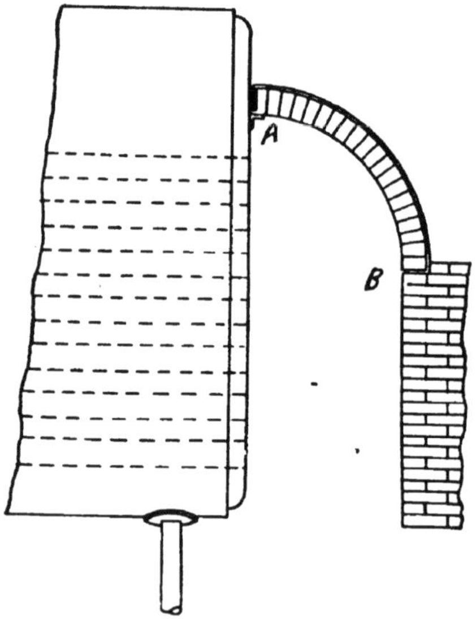

FIG. 52. BACK ARCH IN PLACE.

the lining is given a batter of 3 inches, starting at the level of the grates, then the width of grate will be 6 inches less.

Ques. 228.—What is the ordinary ratio of grate surface to heating surface for land boilers, with natural draught?

Ans.—One square foot of grate surface to every 36 square feet of heating surface.

Ques. 229.—What ratio of grate surface to heating surface is usually chosen with forced draught?

Ans.—One square foot of grate surface to 40 square feet of heating surface, and in some instances the ratio is as high as 1 to 50.

Ques. 230.—How many different styles of grate-bars are in general use?

Ans.—Four; first, the common stationary grate, consisting of a plain cast-iron bar tapered in cross

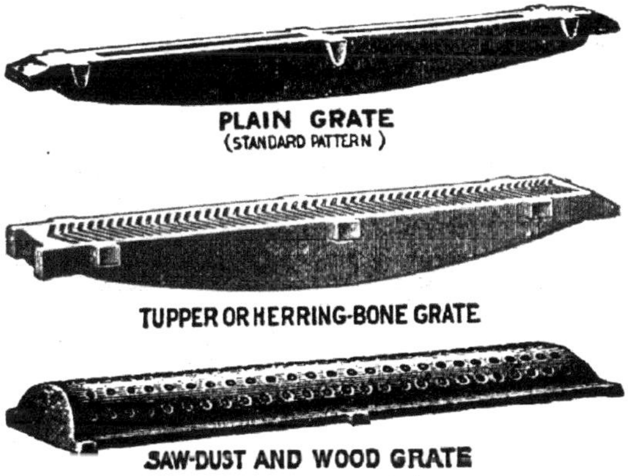

Fig. 53. Grate Bars.

section and having small projections cast on the sides to keep the bars apart a sufficient distance; second, herring-bone grates, consisting of channel-shaped cast-iron bars having V-shaped openings on top to allow the air to pass through to the fire; third, shaking or rocking grates, fourth, dumping grates.

Ques. 231.—What percentage of the total grate area is usually allowed for the admission of air through the grates?

Ans.—From 30 to 50 per cent, depending upon the kind of coal used.

Ques. 232.—What is the heating surface of a boiler?

Ans.—All the surfaces that are in contact with and covered by water on one side and surrounded by flame or hot gases on the other side. The areas of these surfaces are estimated in square feet and added together.

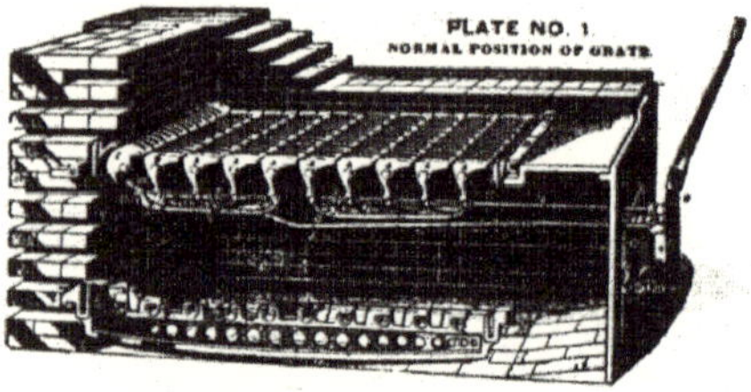

FIG. 54. M'CLAVE'S GRATES.

FIG. 55. M'CLAVE'S GRATES.

Ques. 233.—Is it possible to estimate the horse-power of a boiler from its heating surface?

Ans.—It is in a general way, but not accurately.

Ques. 234.—How many square feet of heating surface are usually allowed per horse-power?

Ans.—From 10 to 16 square feet, depending entirely upon the type of boiler.

Ques. 235.—Give some examples.

Ans.—For water-tube boilers 10 to 12 square feet of heating surface; for horizontal fire-tube, 12, for vertical fire-tube, 12 to 15, and for locomotive boilers, 12 to 16 square feet of heating surface per horse-power.

Ques. 236.—Why this difference?

Ans.—Because the heating surface is more effective in some types of boilers than it is in others.

Ques. 237.—What is the rule for calculating the heating surface of a horizontal tubular boiler?

Ans.—Taking the dimensions in inches, multiply two-thirds of the circumference of the shell by its length. Multiply the inside circumference of one of the tubes by its length, and this product by the number of tubes. Add these two products together, and to this sum add two-thirds of the combined areas of both tube-sheets and from this latter sum subtract twice the combined sectional areas of all the tubes. The result will be the heating surface in square inches, which, divided by 144, will give the number of square feet of heating surface.

Ques. 238.—What is the rule for finding the heating surface of vertical fire-box boilers?

Ans.—Multiply the circumference of the fire-box by its height above the grate. Find the heating surface of the tubes by the process given in the former rule and add these two products together, and to this add the area of the lower tube sheet. From this sum deduct the sectional area of all the tubes. The dimensions having been taken in inches, the result should be divided by 144 to ascertain the number of square feet of heating surface.

Ques. 239.—Why is the inside circumference of the tubes taken?

Ans.—Because in fire-tube boilers this is the portion that is directly exposed to the heat.

Ques. 240.—Why are the combined sectional areas of the tubes subtracted from the area of that portion of the tube-sheets that is exposed to the heat

Ans.—Because the effective heating surface of a tube-sheet is the surface remaining after the areas of the openings through the tubes is deducted.

Ques. 241.—What is implied in the expression "a 3-inch boiler tube?"

Ans.—It means a tube 3 inches in external diameter.

Ques. 242.—Such being the case, which diameter should be considered in calculating the heating surface of fire-tubes?

Ans.—Only the inside diameter, which equals the outside diameter minus twice the thickness of the tube.

Ques. 243.—In calculating the heating surface of the tubes of water-tube boilers which diameter should be taken?

Ans.—The outside diameter, for the reason that the outside circumference is exposed to the heat.

Ques. 244.—How is the heating surface of a water-tube boiler ascertained?

Ans.—Much depends upon the style of boiler. A general rule and one that will apply in all cases, is to multiply the outside circumference of one of the tubes by its length, and this product by the number of tubes that are of a similar length and diameter. If there are various sections of tubes of varying lengths, the heating surface of each section must be ascertained separately and the whole added together. To this sum must be added the combined areas of those portions of the headers that are directly exposed to the heat, having first deducted the sectional area of the tubes. All of those portions of the steam and water-drums that are directly exposed to the heat should be estimated as heating surface also.

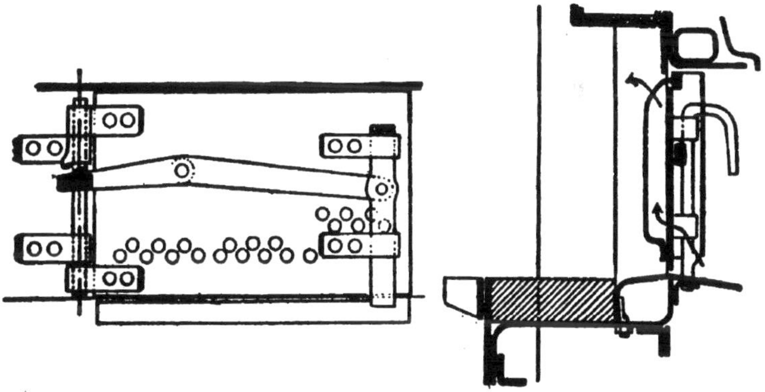

FIG. 56 DOOR OF A BELLEVILLE BOILER.

The door proper has an outer and inner plate, the former being a screen plate with edges open for the admission of air The door is perforated with holes at the lower part, through which the air is drawn, and the inner plate, which is of cast iron, is closed at the bottom, and has holes for the discharge of air at the top. When the fires are alight, there is a continuous current of air flowing into the furnace through these plates.

Ques. 245.—What is the rule for ascertaining the heating surface of a Scotch boiler?

Ans.—The grates being set in the large main flues, only one-half of each flue area is available as heating surface. The following rule applies: To one-half the combined area of the main flue add the area of one head between the grate and water-line, minus the total cross-section of the tubes, plus one-half the cross-section of main flues, plus the combined inside area of the tubes, plus the inside area of the combustion chamber.

Ques. 246.—Give the rule for finding the heating surface of a corrugated flue.

Ans.—Multiply the average inside diameter in feet by the length of the flue in feet, and this product by the

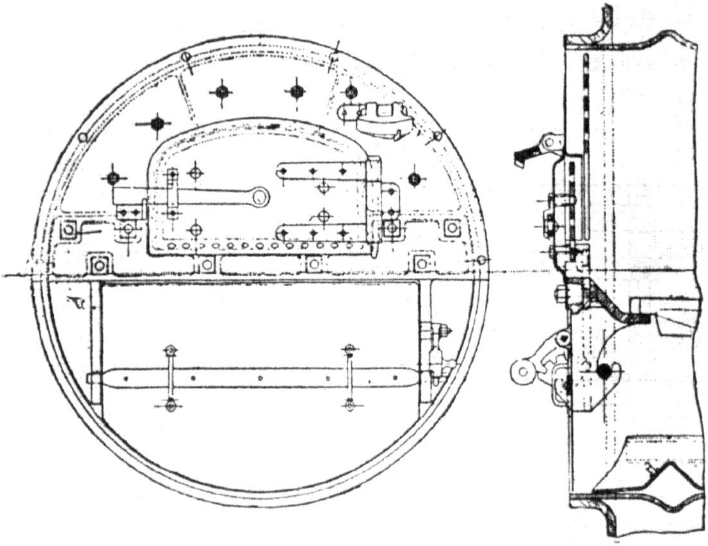

Fig. 57.

Shows another variety, the air being admitted through holes at the bottom of the wrought-steel door proper, a perforated inner cast-iron plate being fitted to shield the door. The wrought-steel furnace frame which carries the door also has an inner shield plate of cast-iron perforated with holes.

constant 4.93. The result is square feet of heating surface.

Ques. 247.—What is the duty of a safety valve?

Ans.—To automatically relieve the boiler of all pressure above a certain prescribed working pressure by allowing the surplus steam to escape into the atmosphere.

Ques. 248.—If a boiler had no safety valve, or if the

safety valve should refuse to work, and all other exit from the boiler be closed, and heat continuously applied, what would be the result?

Ans.—An explosion must of necessity occur.

Fig. 58. Pop Valve.

Ques. 249.—How many types of safety valves are in use?

Ans.—Two; the lever safety valve, and the spring-pop safety valve.

Ques. 250.—Which is the best adapted to all kinds of service?

Ans.—The spring-loaded pop safety valve is, for the

reason that any inclination of the boiler, such as that caused by the vessel's pitching and rolling in a heavy sea, does not interfere with the working of a spring-pop valve,

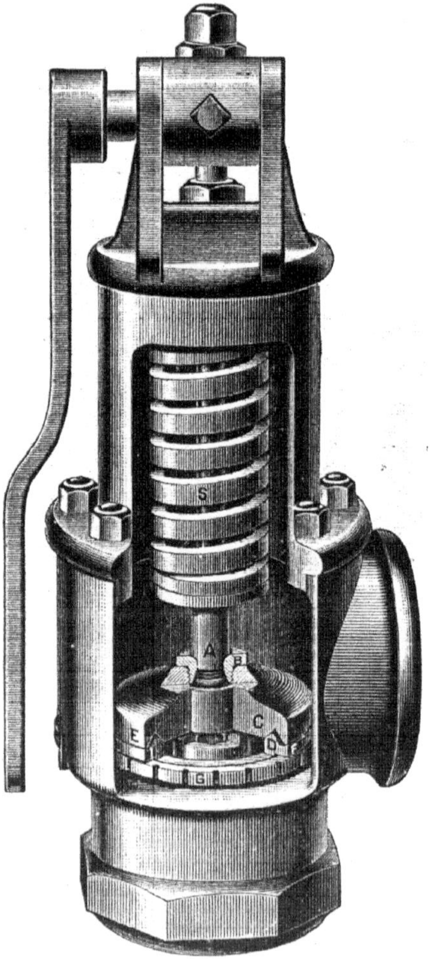

FIG. 59. INSIDE VIEW OF A POP SAFETY VALVE.

while on the other hand the leverage of a weighted lever valve decreases with any inclination of the boiler that would momentarily put the lever in an inclined position.

Ques. 251.—What is the United States marine rule for determining the area of lever safety valves for boilers?

Ans.—"Lever safety valves to be attached to marine boilers shall have an area of not less than 1 square inch to

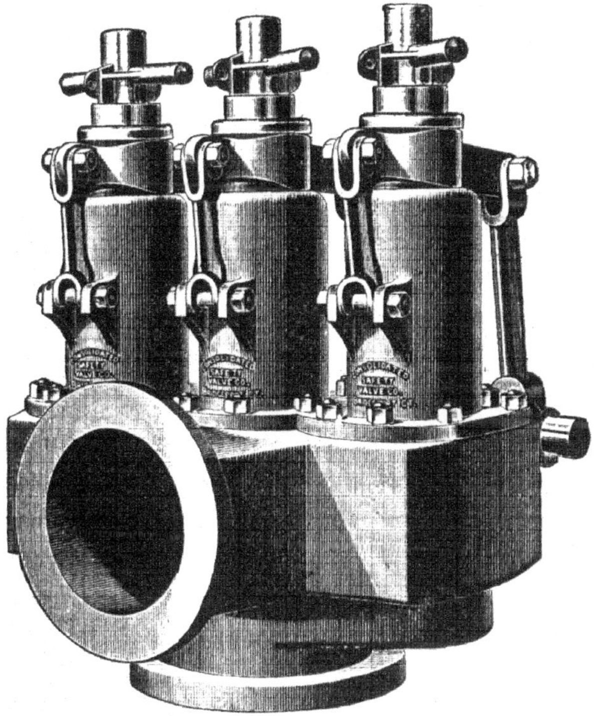

FIG. 60. TRIPLEX POP SAFETY VALVE.

every 2 square feet of grate surface in the boiler, and the seats of all such safety valves shall have an angle of inclination of 45 degrees to the center line of their axis."

Ques. 252.—What is the rule regarding spring pop safety valves?

Ans.—Three square feet of grate surface are allowed to each square inch of safety-valve area.

Ques. 253.—What other and more reliable method is there of calculating safety-valve area?

Ans.—The method by which the area of the valve is based upon the quantity of steam that the boiler is capable of generating.

Ques. 254.—Why is this method more reliable?

Ans.—For the reason that the rate of combustion varies greatly under different conditions, as, for instance, when forced draught is employed, a much higher rate of combustion is attained than is possible with natural draught.

Ques. 255.—Do the standard rules given in answers 251 and 252 hold good for safety-valve areas for all pressures?

Ans.—No; because the rate of efflux for steam increases as the pressure increases. Therefore, for the higher pressures the total safety-valve area may be reduced.

Ques. 256.—What should be the lift of a safety valve in order to allow the proper area of escape?

Ans.—One-fourth of the diameter of valve.

Ques. 257.—What is the rule for ascertaining the pressure at which a lever safety valve will lift when the weight and its distance from the fulcrum are known, as also the effective weight of the valve, stem, and lever?

Ans.—Multiply the weight by its distance from the fulcrum. Multiply the weight of the valve and lever by the distance of the stem from the fulcrum, and add to the former product. Divide the sum of the two products by the product of the area of the valve multiplied by its distance from the fulcrum. The result will be the pressure in pounds at which the valve will lift.

Ques. 258.—What is the rule for finding the distance that the weight should be placed from the fulcrum for a required pressure?

Ans.—Multiply the area of the valve by the pressure

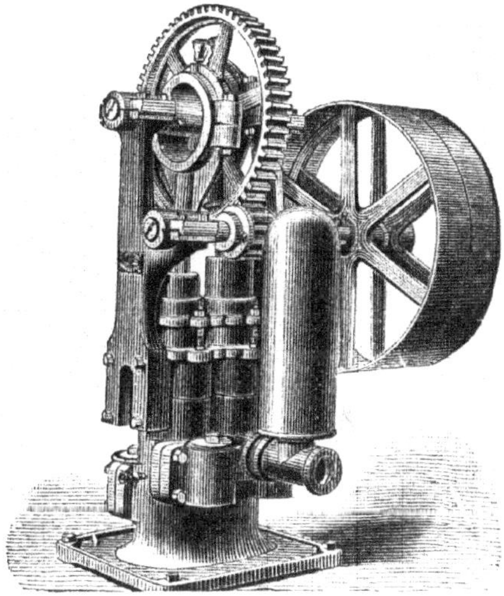

FIG. 61. DAVIS BELT DRIVEN FEED PUMP.

at which it is desired to have it lift, and from this product subtract the effective weight of the valve and lever. Multiply the remainder by the distance of the stem from the fulcrum, and divide by the weight. The quotient will be the required distance.

Ques. 259.—What is the rule for ascertaining the weight required when all of the other factors are known?

Ans.—Multiply the area of the valve by the pressure, and from the product deduct the effective weight of the valve and lever. Multiply the remainder by the distance of the stem from the fulcrum and divide by the distance of the ball or weight from the fulcrum. The quotient will be the required weight in pounds.

FIG. 62. PHANTOM VIEW OF MARSH INDEPENDENT STEAM PUMP.

Ques. 260.—What can be said in general regarding the safety valve?

Ans.—It is one of the most useful and important adjuncts of a steam boiler, and if neglected, serious results are apt to follow.

Ques. 261.—Mention the two standard methods of supplying the feed-water to boilers under pressure?

Ans.—First, by the feed-pump; second, by the injector.

Ques. 262.—What advantage has the feed-pump over the injector?

Ans.—The advantage of being able to draw its supply of water from a heater, in which the exhaust steam is utilized for heating the feed-water before it enters the boiler. Great economy in fuel is thereby effected.

Ques. 263.—What is a duplex pump?

Ans.—A duplex pump consists of two steam-cylinders and two water-cylinders, each having the necessary pistons and valves. The steam-valves of one side are

FIG. 63. WORTHINGTON DUPLEX BOILER FEED PUMP.

operated by the other side, and vice versa. Both water cylinders discharge into the same main. A common suction main serves both water-cylinders also.

Ques. 264.—If one side of a duplex pump becomes disabled from any cause, how may the other side be operated for the time being?

Ans.—Loosen the nuts or tappets on the valve-stem of the broken side and place them far enough apart so that the steam-valve will be moved through only a small portion of its stroke, thereby admitting only steam enough to move the empty steam-piston and rod, and thus work the steam-valve of the remaining side. The packing on the piston-rod of the broken side should be screwed up tightly, so as to create as much friction as possible, there being no resistance in the water end. In this manner the pump may be operated for several days or weeks, and thus prevent a shut-down.

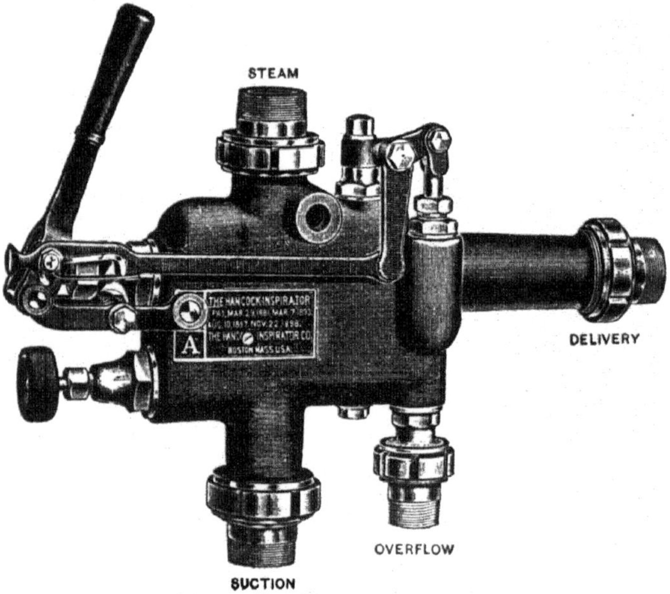

FIG. 64. THE HANCOCK INSPIRATOR.

Ques. 265.—How is the velocity of flow, or piston-speed per minute of a pump ascertained?

Ans.—Multiply the number of strokes per minute by the length of stroke in feet, or fractions thereof. This will give the piston-speed in feet per minute.

Ques. 266.—How is the velocity of flow in the discharge-pipe ascertained?

Ans.—Divide the square of the diameter of the water-cylinder in inches by the square of the diameter of the discharge-pipe in inches, and multiply the quotient thus

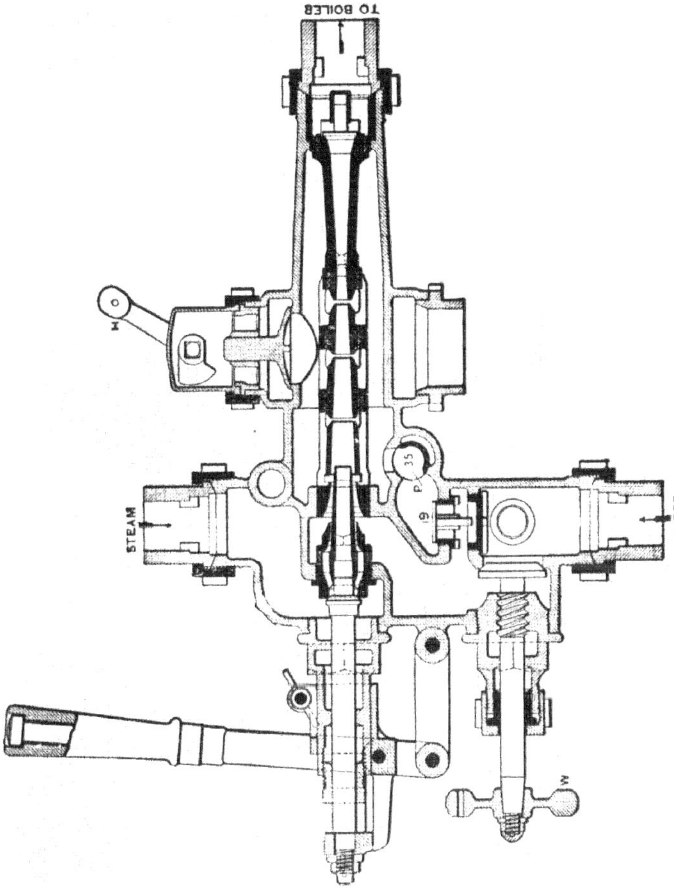

FIG. 65. THE SIMPLEX INJECTOR.

obtained by the piston-speed in feet per minute of the pump.

Ques. 267.—When the velocity in feet per minute is

known, how may the number of cubic feet discharged per minute be ascertained?

Ans.—Multiply the area of the pipe in square inches by the velocity in feet per minute, and divide by the constant 144. The result will be the number of cubic feet of water or other fluid discharged per minute.

Ques. 268.—How may the required size and capacity of feed-pump for a certain boiler be ascertained?

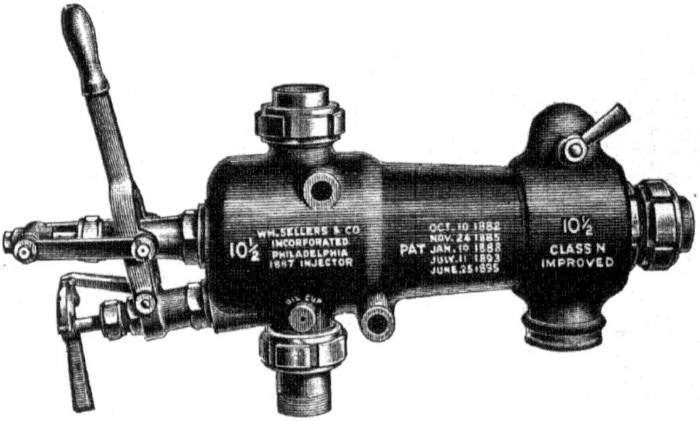

FIG. 66. THE SELF-ACTING INJECTOR

Ans.—Multiply the number of square feet of grate surface by the number of pounds of coal it is desired to burn per hour per square foot of grate. This will give the total coal consumed per hour, which, multiplied by the number of pounds water evaporated per pound of coal will result in the total number of pounds water required per hour.

Ques. 269.—How may the required size of the feed-pump be ascertained from the number of square feet of heating surface?

Ans.—Allow a pump capacity of 1 cubic foot of water per hour for each 15 square feet of heating surface.

Ques. 270.—How can an injector lift and force water

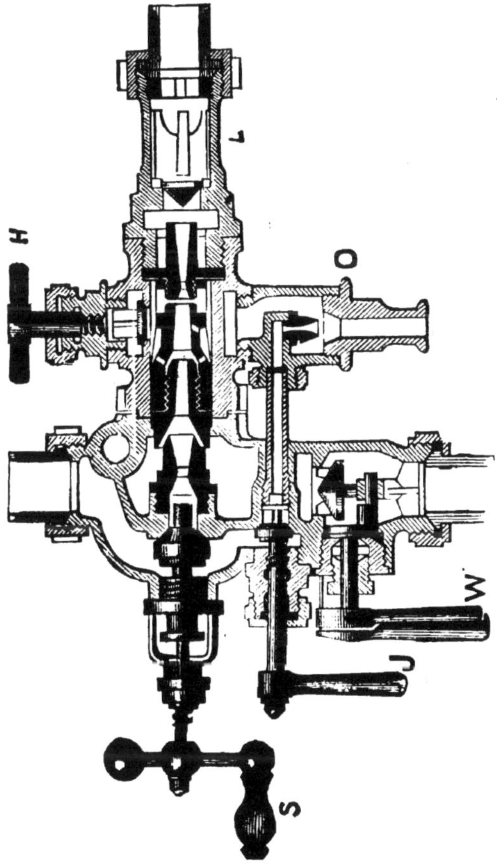

Fig. 67. The "Monitor" Injector, Interior View.

into the boiler against the same or even higher pressure than the pressure of the steam supplied to the injector?

Ans.—An injector works because the steam imparts sufficient velocity to the water to overcome the pressure in the boiler.

Ques. 271.—What is the velocity of a jet of steam under 180 pounds pressure issuing from a nozzle?

Ans.—About 3,600 feet per second.

Ques. 272.—What is the velocity of a jet of water under a pressure of 180 pounds issuing from a nozzle?

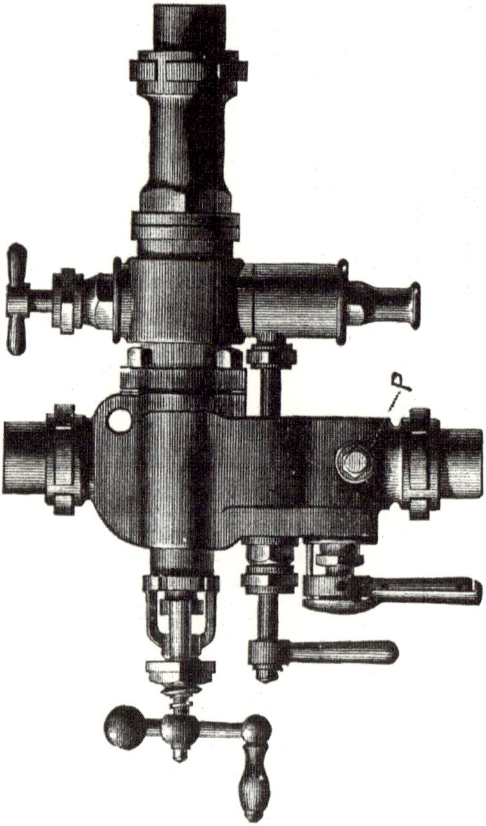

Fig. 68. The "Monitor" Injector, Exterior View.

Ans.—Only 164 feet per second.

Ques. 273.—Why does the steam have so much greater velocity than the water, when the pressure in both instances is the same?

Ans.—Because of the latent heat that is stored in the steam.

Ques. 274.—What is the purpose of the combining tube in an injector?

Ans.—To bring the jet of steam and the jet of water into close contact in order that the steam may be con-

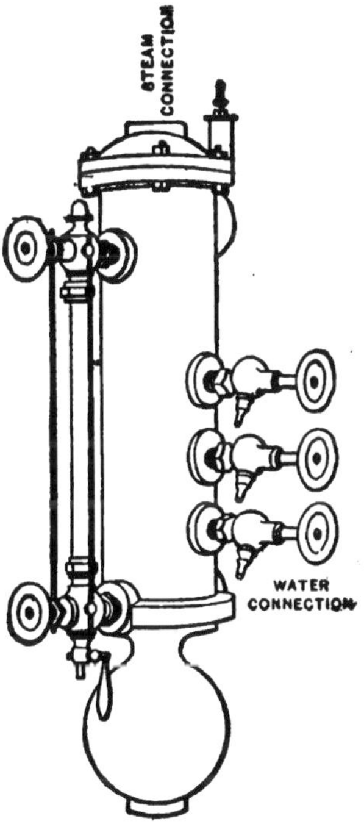

FIG. 69. WATER COLUMN.

densed and the size of the jet reduced sufficiently to allow it to enter the delivery tube, which is of smaller diameter than the combining tube.

Ques. 275.—What is the velocity of the combined jet of water and condensed steam as it leaves the combining tube and enters the delivery tube, assuming the steam-pressure in the boiler to be 180 pounds per square inch?

Ans.—198 feet per second.

Ques. 276.—What velocity is actually needed to cause the jet to enter the water-space of the boiler carrying 180 pounds pressure?

Ans.—Only 164 feet per second. The excess of 34 feet per second imparted to the velocity of the jet serves to overcome the friction of the feed-pipe and the resistance of the main check-valve.

Ques. 277.—In general terms, then, to what is the action of the injector due?

Ans.—The action of the injector is due to the high velocity with which a jet of steam strikes the water entering the combining tube, imparting to it its momentum and forming with it during condensation a continuous jet of smaller diameter, having sufficient velocity to overcome the pressure in the boiler.

Ques. 278.—What is the object in fitting a boiler with a check-valve in the feed-pipe?

Ans.—A check-valve is for the purpose of preventing the water in the boiler from backing up into the feed main and feed-pump.

Ques. 279.—Where should the check-valve be located?

Ans.—In the feed-pipe, as near to the boiler as possible.

Ques. 280.—For what purpose are gauge-cocks and water-gauge glasses?

Ans.—They are for the purpose of indicating the height of the water in the boiler while it is under pressure.

Ques. 281.—Describe the construction and operation of a glass water-gauge?

Ans.—A water-gauge, otherwise known as a water column or combination, is a cast-iron or brass cylinder connected to the steam-space of the boiler at the top, and to the water-space near the bottom. The normal position of the safe water-level is near the middle of the water-

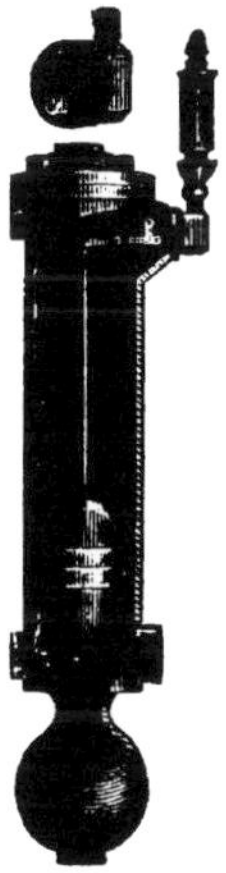

FIG. 70. LOW WATER ALARM.

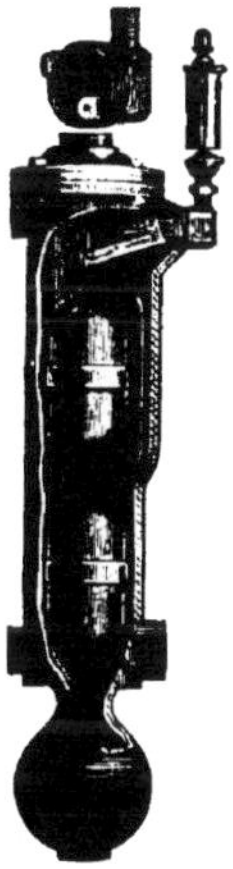

FIG. 71. COMBINED HIGH AND LOW WATER ALARM.

column, into one side of which are screwed brass fittings for the glass tube or water-glass, which is a strong tube of special manufacture. Each end of this tube passes through a stuffing box in the brass fittings. The joint is made steam tight by a rubber ring that fits around the tube and is compressed by a follower screwed onto it. The fittings that connect the water-column with the boiler are, or at least should be, equipped with automatically closing ball valves which will act in case the gauge-glass breaks.

Ques. 282.—Where are the gauge-cocks or test-cocks usually connected?

Ans.—They are usually connected to the water-column cylinder in such a position that the lowest one is at the desired water-level, one a few inches above that, and the third near the highest point of the heating service. These test-cocks should be opened several times a day in order to keep them clear for use in case the gauge-glass breaks.

Ques. 283.—What is liable to happen to the water-column?

Ans.—Unless the water and sediment are frequently blown out of it through the valve at the bottom provided for this purpose, the tubes and connections will become clogged, thus preventing a free circulation of the water, and the true water-level in the boiler will not be indicated as it should be.

Ques. 284.—What is a fusible plug?

Ans.—A fusible plug is a 1-inch brass pipe threaded plug, having its center drilled out to a diameter of not less than ½ inch, and the hole filled with Banca tin or other fusible metal.

Ques. 285.—Where should a fusible plug be attached to a boiler?

Ans.—A fusible plug should always be attached to that portion of the boiler that is first liable to become overheated on account of the water-level becoming too low.

Ques. 286.—Mention some proper locations for fusible plugs in various types of boilers?

Ans.—The back head of a horizontal tubular boiler,

about 3 inches above the top row of tubes, the crown-sheet of a horizontal fire-box boiler; the lower tube-sheet of a vertical boiler, or sometimes in one of the tubes a few inches above the tube-sheet; in the lower side of the upper drum of a water-tube boiler. The fusible metal which fills the center of the plug is of conical form in order to prevent its being blown out by the pressure behind it. On the other hand, the melting point of this fusible metal is such that when the water falls below it, and the steam under pressure in the boiler comes in contact with it, the metal is melted and runs out, thus allowing the steam to escape through the hole and give the alarm. If the melted plug is located in the crown-sheet of a fire-box boiler, the escaping steam and water will quench the fire and thus lessen the danger of burning the sheet.

FIG. 72.

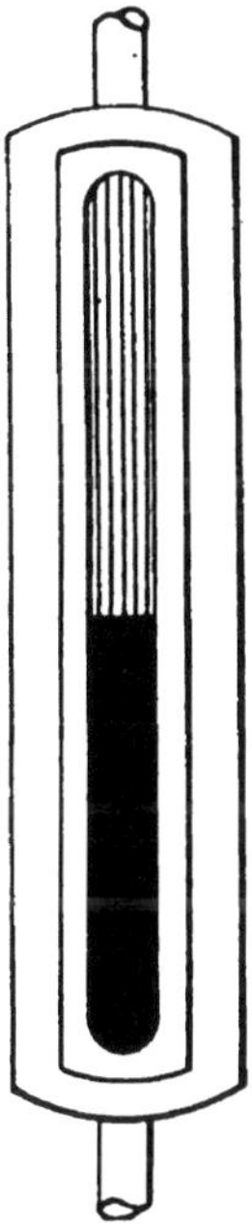

FIG. 72a.

Klinger's Water Gauge Mounting.—The usual round thin gauge glasses give trouble with high-pressure steam, owing to frequent fractures, while the water level is often indistinct. Klinger's glass, designed to obviate these defects, gives promise of success. It consists of a thick flat glass, with smooth front and serrated back, shown in section Fig. 72. A and B, the front and back of the mounting, are bolted together with the glass and packing, shown by thick lines, between them. The serrations, when clean, cause the water to appear black, as in Fig. 72a.

Ques. 287.—For what purpose is a steam-gauge attached to a boiler?

Ans.—For the purpose of indicating the number of pounds pressure per square inch in the boiler.

Ques. 288.—What type of steam-gauge is in most general use?

Ans.—The Bourdon spring tube gauge.

Ques. 289.—Describe the construction of this gauge, and the principle upon which it operates?

Ans.—The Bourdon gauge consists of a thin, curved,

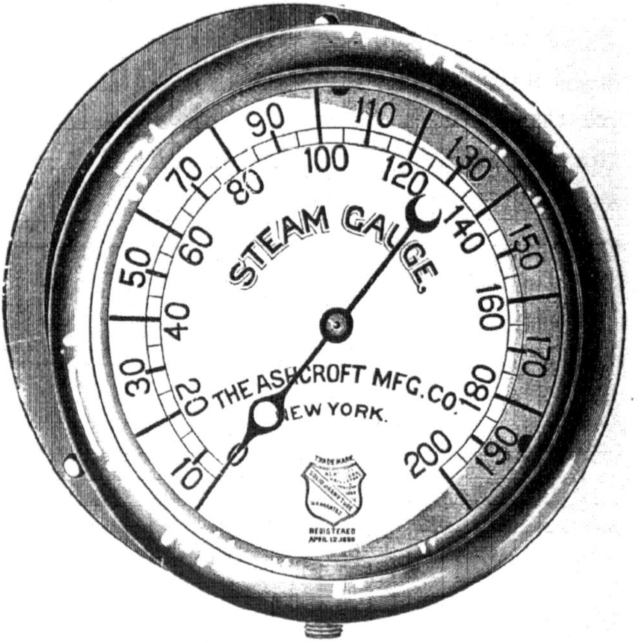

FIG. 73. AUXILIARY SPRING PRESSURE GAUGE.

flattened metallic tube closed at both ends and connected to the steam-space of the boiler by a small pipe bent at some portion of its length into a curve or circle that becomes filled with water of condensation, and thus prevents the live steam from coming directly in contact with the tube or spring, while at the same time the full pressure of steam in the boiler acts upon the tube, tending to straighten it. The end or ends of the spring tube being free to move, and connected by a suitable geared rack and pinion with the pointer of the gauge, causes it

FIG. 74. AUXILIARY SPRING PRESSURE GAUGE, SECTIONAL VIEW.

to move across the face of the dial, thus indicating the pressure of the steam in pounds per square inch on the inner surface of the boiler. When there is no pressure in the boiler the pointer should stand at 0.

Ques. 290.—How should steam-gauges be cared for?

Ans.—They should be tested frequently by comparing them with a gauge that has been tested against a column of mercury.

Ques. 291.—How should the steam-space of the boiler be connected to the main steam-pipe or header?

Ans.—There should be a steam stop-valve placed in the connection between the boiler and the header. The valve

FIG. 75. SECTIONAL VIEW AMERICAN PRESSURE GAUGE.

used for this purpose is usually an angle-valve, and should be constructed so as to close automatically, especially in a battery of two or more boilers.

Ques. 292.—Why should this valve be self-closing in case the pressure in the header is higher than the pressure in the boiler?

Ans.—In order that in case of an accident to one of a battery of boilers the steam may be prevented from passing out of the header and into the disabled boiler.

Ques. 293.—Describe the construction and operation of an automatic steam stop-valve.

Ans.—The valve is opened and closed by means of a screw-stem passing out through the stuffing box, and fitted with a hand-wheel outside. In large-size valves this screw-thread is carried in a strong yoke outside the casing. The pressure from the boiler is on the under side of the valve-disk, thus tending to open it. The stem or spindle is independent of the valve, and is hollow to allow a smaller size sliding spindle connected to the valve to pass into it. This spindle serves to guide and hold the valve steady, while at the same time the valve is free to close automatically any time that the pressure in the main exceeds the pressure in the boiler.

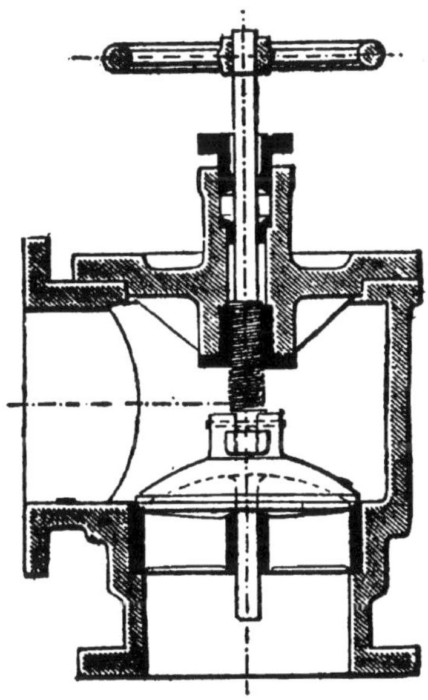

FIG. 76. SECTION OF AN ANGLE STOP-VALVE.

Ques. 294.—How is the steam admitted to the whistle or the steam siren?

Ans.—Through a special stop-valve, usually of the self-closing type, being worked by a spring on the valve.

Ques. 295.—Describe the action of the steam whistle.

Ans.—The steam whistle produces its sound by the vibrations of a thin stationary metallic cylinder, under the impact of the steam.

Ques. 296.—How does the steam siren produce its sound?

Ans.—By means of the rotations of a small slotted wheel which in turning opens and closes narrow slots in the casing.

Ques. 297.—How may the passage of water from the boiler into the steam-pipe be prevented to a large extent?

Ans.—By means of an internal pipe-extension called a dry pipe, that collects the steam from all parts of the steam-space through narrow slots on its upper side. The shape of these slots has a straining action on the steam.

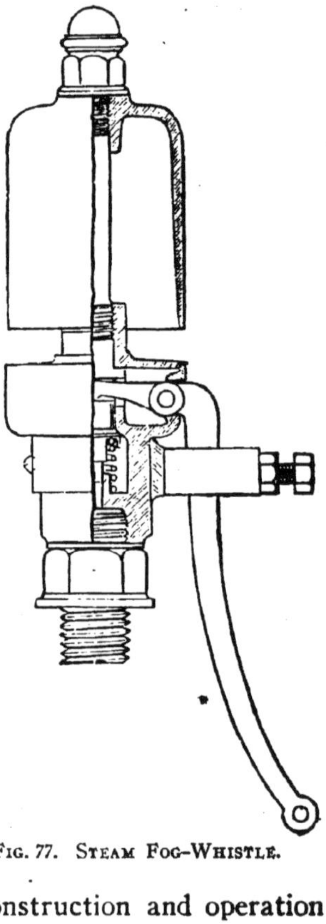

FIG. 77. STEAM FOG-WHISTLE.

Ques. 298.—What is the object in equipping a boiler with a surface blow-off?

Ans.—In order that it may catch and pass off impurities, such as grease, oil, and scum, floating on the surface of the water.

Ques. 299.—Describe the construction and operation of the surface blow-off.

Ans.—It is connected to the boiler near the water-level, and carries an internal pipe-extension that ends in a flat pan, directly below the water-line. It should be opened quite frequently, especially when muddy water is being fed to the boiler. This will allow the accumulated scum to pass out.

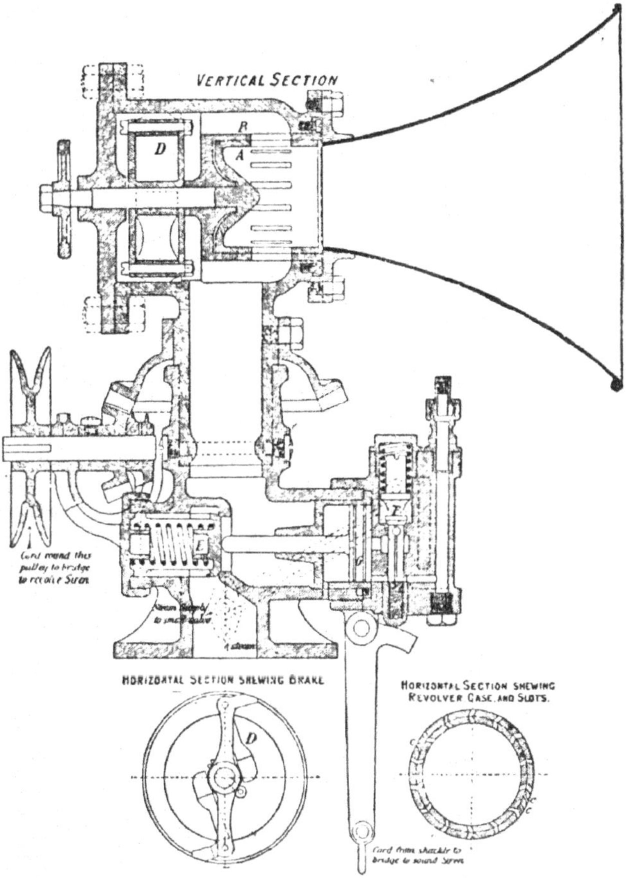

FIG. 78. SECTION OF A STEAM SIREN.

Ques. 300.—Where and how should the bottom blow-off be connected?

Ans.—The bottom blow-off should be connected to the lowest section of the boiler, and should be fitted with a straight-way valve, or a cock, in order that there may be no obstruction to the free passage of the mud and other sediment when the boiler is being cleaned.

Ques. 301.—For what purpose is the hydrometer-cock, and where is it located?

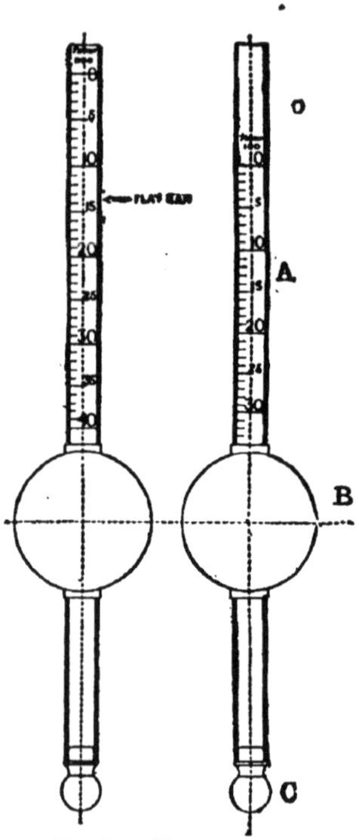

FIG. 79. HYDROMETER.

Ans.—In the marine service the water used in the boilers is more or less impregnated with solid matter, and it becomes necessary to test the density of the water in the boilers at certain intervals. The hydrometer-cock is for the purpose of drawing off a quantity of water from the boiler for testing, and is fitted to the water-space of the boiler.

Ques. 302.—Describe the construction and use of the hydrometer.

Ans.—It is an instrument having a long, slender stem, made of either glass or metal. There are two bulbs in the stem. The smaller one is loaded and the larger one is hollow and filled with air, which gives the instrument buoyancy, and keeps it in a vertical position. The stem is graduated in degrees, each degree representing the presence of one-tenth the solid matter in sea-water.

Ques. 303.—What proportion of sea-water is solid matter?

Ans.—One thirty-second part.

Ques. 304.—Upon what principle are the readings taken from the hydrometer based?

Ans.—Upon the principle that when any body floats freely, the weight of the liquid displaced is equal to the weight of the body floating, so that the higher the density of the liquid the less depth will the body sink in it. If the instrument sinks only to the zero mark on the scale, the water is fresh: if it sinks to 10 degrees, it indicates the presence of one-thirty-second part of solid matter, and if it sinks to 40 degrees, it indicates a density caused by the presence of four times as much solid matter as there is in sea-water.

Ques. 305.—How is the water in the boiler tested with the hydrometer?

Ans.—A quantity of water is drawn off through the hydrometer-cock, fitted for this purpose into a long pot, into which the instrument is inserted.

Ques. 306.—How are boiler hydrometers graduated, with reference to temperature?

Ans.—They are usually graduated to suit a temperature of 200 degrees Fahrenheit, as that is about the temperature of the water a few seconds after being drawn off for testing.

Ques. 307.—How are the expansion and contraction of steam-pipes provided for?

Ans.—In the smaller sized pipes a bend can be put in the length of pipe that will answer the purpose, but in the large pipes an expansion joint, having a stuffing box for the pipe to slide in and out of the adjacent pipe is fitted.

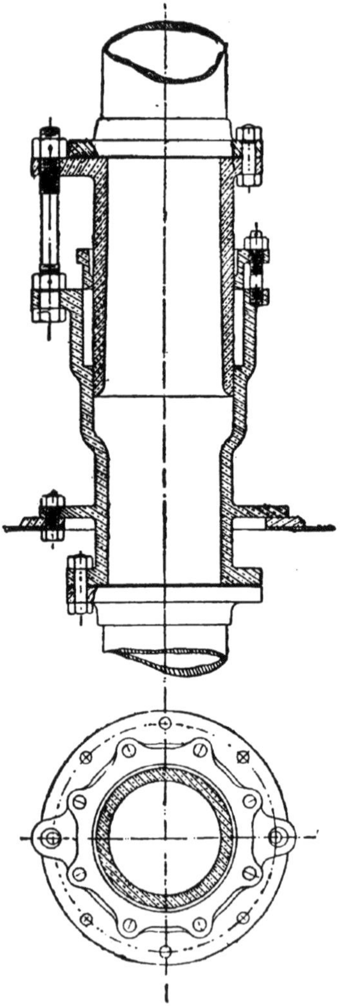

FIG. 80. EXPANSION JOINT.

Ques. 308.—Why is it necessary to place a separator in the line of pipe leading from the boiler to the engine?

Ans.—The object of a separator is to provide an additional safeguard against priming, by preventing any water in the steam-pipe from entering the cylinder.

Ques. 309.—Describe the ordinary separator.

Ans.—It is a metal cylinder larger in diameter than the steam-pipe, and connected to the pipe near the engine, by flange connections in such a manner that the larger portion of the separator hangs in a vertical position below the pipe. It is divided from the top nearly to the bottom by a diaphragm, and the steam enters on one side, near to the top, and impinges against the diaphragm, passes underneath it, and out on the other side near the top.

Any water that reaches the separator is mostly left at the bottom, only the steam passing on to the engine cylinder. A valve is provided at the bottom of the separator for drawing off the water. The height of the water in the separator is shown by a glass gauge.

Ques. 310.—Describe the automatic steam separator.

Ans.—In addition to the usual diaphragm, it is fitted with an automatic blow-out apparatus, having a float that is raised as the water accumulates, and which by a system of levers opens a valve of large area for drainage. The automatic separation also has a hand blow-off valve.

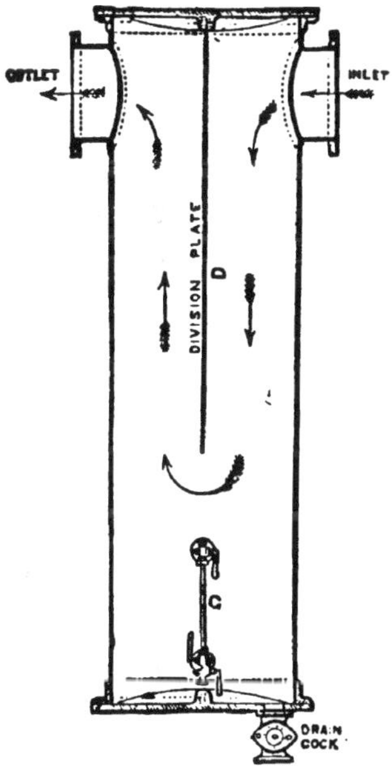

FIG. 81. SEPARATOR.

Ques. 311.—What is an asbestos-packed cock, and where is it used?

Ans.—An asbestos-packed cock has its top and bottom glands packed with asbestos, while the shell also has longitudinal grooves found in it which are packed with asbestos. These cocks are very suitable to use on boilers and steam piping where high pressures are carried, and at locations where cocks are more convenient than valves would be.

Ques. 312.—What are funnel dampers, and for what purpose are they attached?

Ans.—They are hinged dampers fitted in the uptakes leading from the boilers to the funnel, in order that each

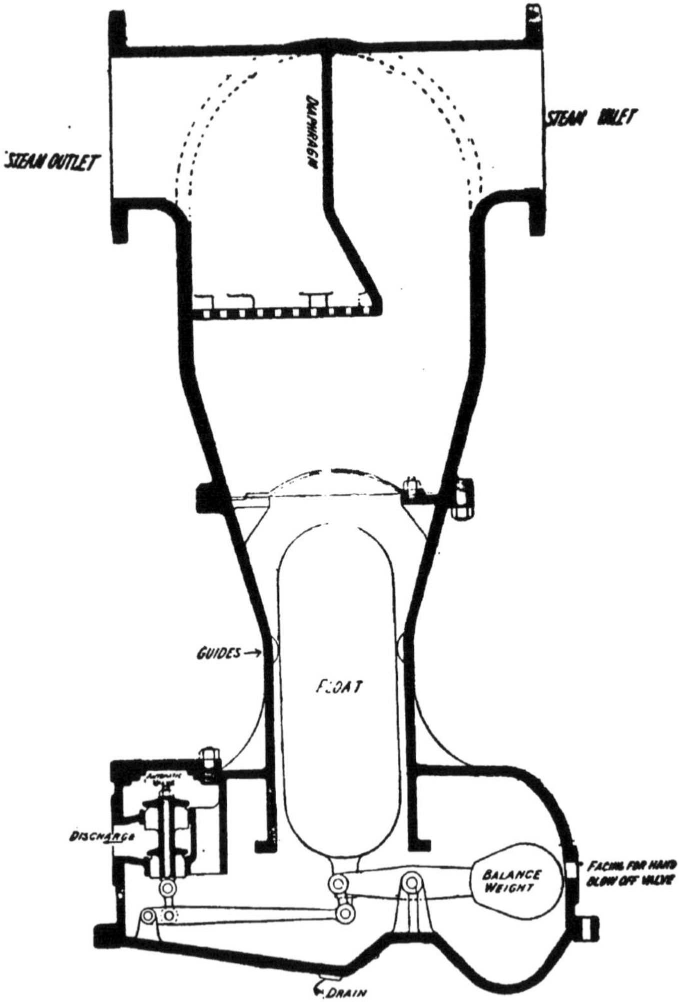

FIG. 82. AUTOMATIC SEPARATOR.

boiler may be shut off from the draught when not in use, and they are also for use when the fires are being cleaned.

These dampers should be fitted so that there are no means of closing them permanently, but that if released they will at once assume the open position.

Ques. 313.—What are funnel stays?

Ans.—Wire ropes carried from the top of the funnel to the ship's sides, and fitted with adjusting screws for the purpose of regulating the strains.

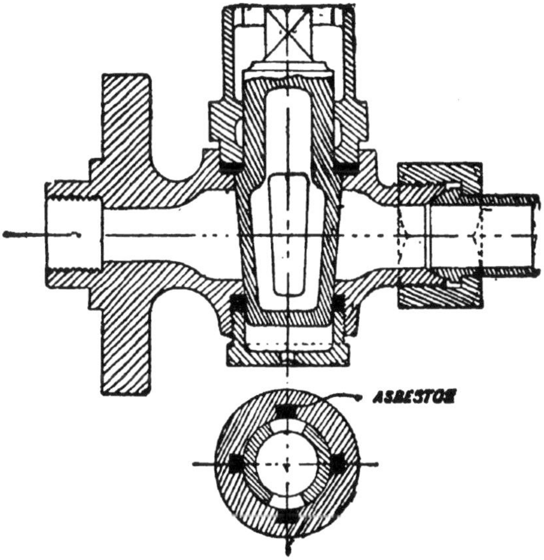

FIG. 83. ASBESTOS-PACKED COCK.

Ques. 314.—What precautions should be taken with these stays before raising steam in the boilers?

Ans.—The adjusting screws should be slackened in order to allow for the expansion in the length of the funnel as it becomes heated.

Ques. 315.—What is the usual height of the funnels of modern vessels?

Ans.—Ninety to 100 feet, measured from the furnaces.

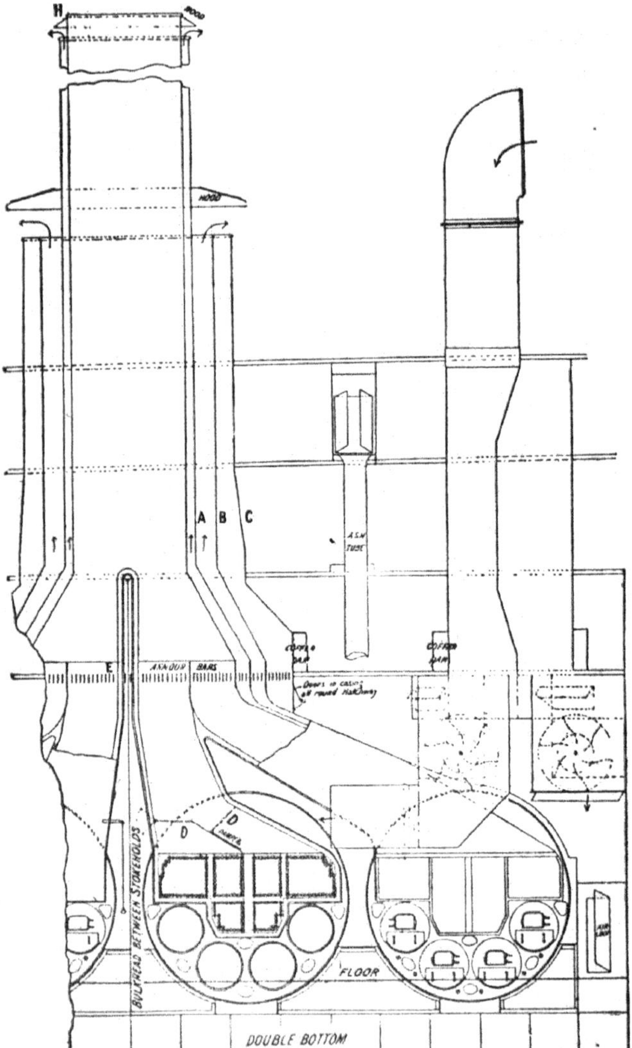

FIG. 84. SECTION OF ARMORED CRUISER, SHOWING STOKE-HOLD AND FUNNELS.

Ques. 316.—For what purpose is the funnel cover?

Ans.—It is fitted over the top of the funnel for use

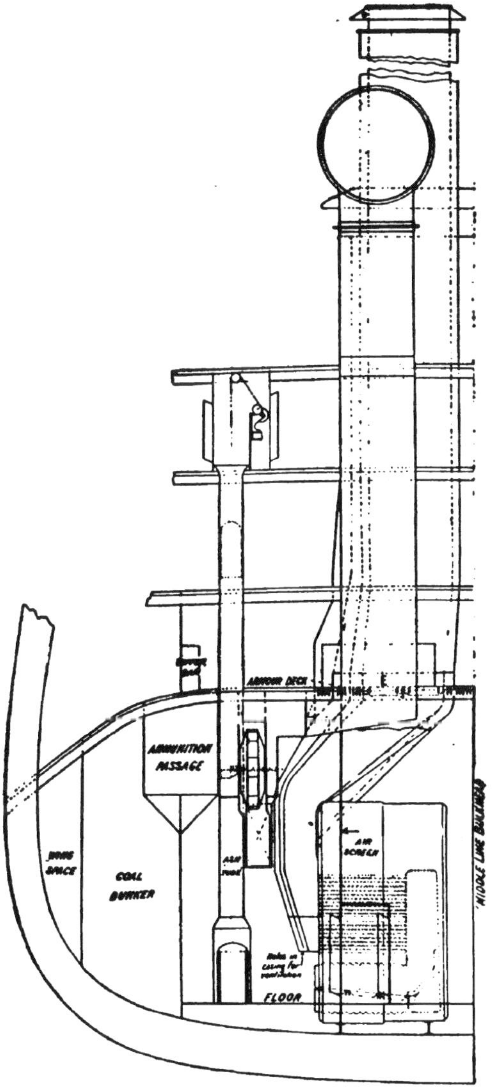

FIG. 85. SECTION OF ARMORED CRUISER, SHOWING AIR SCREEN AND COAL BUNKER.

when the ship is in harbor, or if any of the funnels are not in use, in order to prevent rain-water from entering and corroding the uptakes. These covers are kept a little above the top of the funnel, in order to allow sufficient space for the escape of smoke from small fires used for airing and warming the boilers while they are lying idle.

Ques. 317.—How is the stoke-hold of a steamer ventilated?

Ans.—When natural draught only is used, screens are required to keep the downward current of cool air separate from the upward current of warm or vitiated air, otherwise the circulation will not be as good as it should be.

Ques. 318.—When forced draught is employed for the furnaces, how is the air supplied?

Ans.—One of the oldest and at the same time most expensive methods is to admit a jet of high-pressure steam directly from the boilers to the base of the funnel. This is known as the steam blast. Another plan of using the steam blast is to admit small jets of steam into the furnace, over the fire.

Ques. 319.—What other principal plans for creating forced draught are employed?

Ans.—First, admitting jets of compressed air into the base of the funnel, in a manner similar to the steam-jet; second, fitting a centrifugal fan in the uptake; third, blowing the air into closed ash-pits; fourth, closing the stoke-hold and keeping it filled with slightly compressed air.

Ques. 320.—Of the plans just mentioned, which one is probably the most efficient?

Ans.—Closed stoke-holds, although the third plan, viz., blowing the air into closed ash-pits, is an efficient method, but a certain degree of danger attaches to it, on account of the pressure in the furnaces being greater

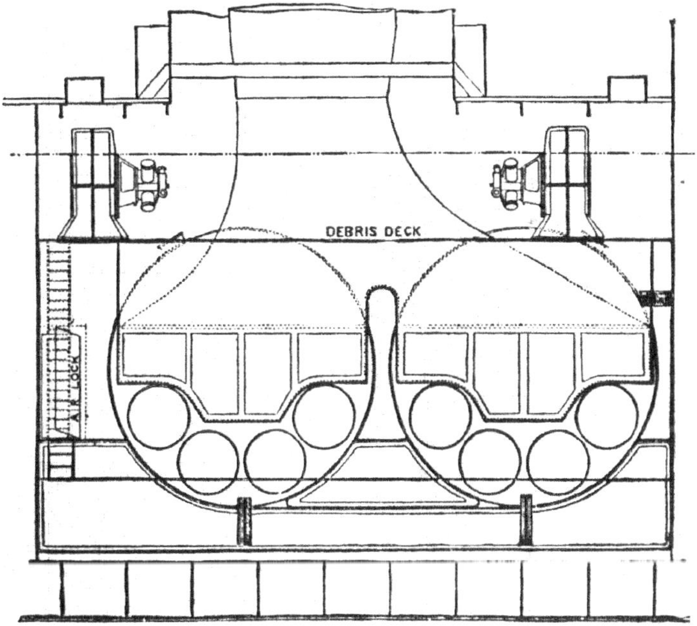

FIG. 86. CROSS SECTION OF STOKE-HOLD, SHOWING AIR LOCK.

than that in the stoke-hold, and unless proper precautions are taken before opening the furnace doors for the purpose of replenishing the fires, the flames may be blown into the stoke-hold and serious results follow.

Ques. 321.—Is this latter system of closed ash-pits much in vogue?

Ans.—It is used to a large extent in the United States navy, also many ships of the mercantile marine service. The British and other navies also use it to some extent.

Ques. 322.—How may this system of creating a forced draught be made safe, so as to guard against the flame being blown into the stoke-hold?

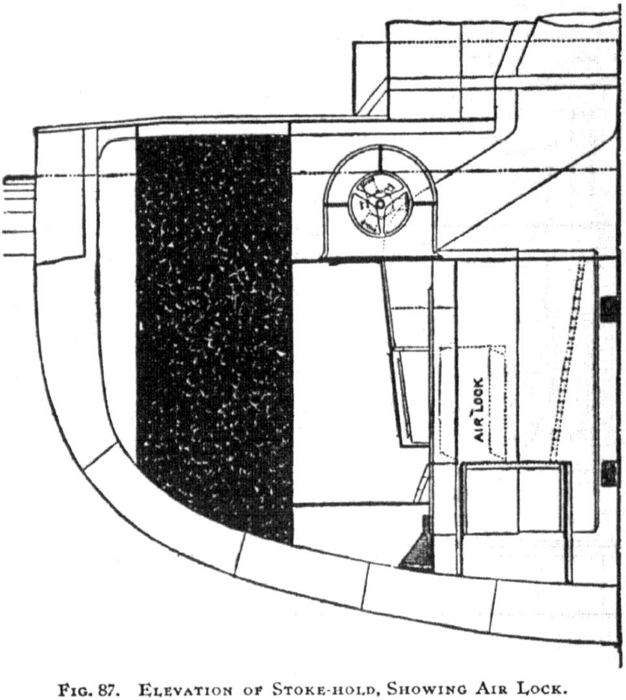

Fig. 87. Elevation of Stoke-hold, Showing Air Lock.

Ans.—By fitting a device that automatically closes the air-supply to the ash-pit when the furnace door is opened for firing.

Ques. 323.—What is the object of providing air-locks in the hold of a vessel?

Ans.—In order to provide for passage to and from the stoke-holds, when under pressure.

Ques. 324.—Describe the construction and operation of an air-lock?

Ans.—An air-lock consists of a small air-tight chamber fitted with two hinged doors opening against the air

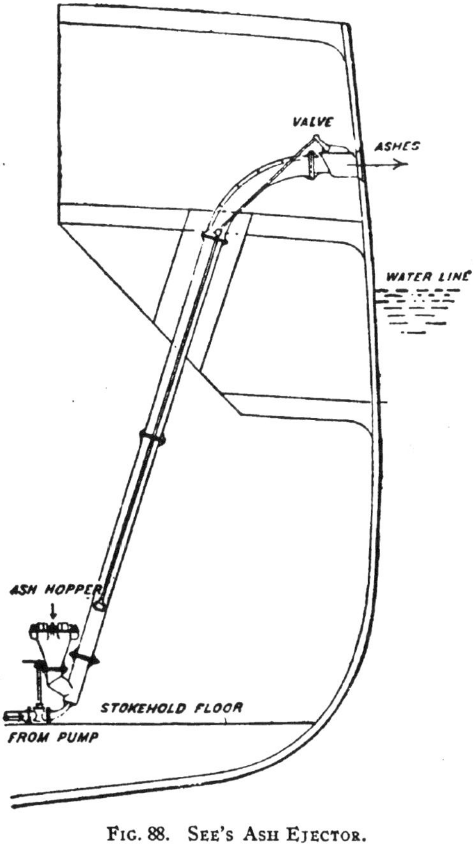

FIG. 88. SEE'S ASH EJECTOR.

In this apparatus, which is fitted in many large passenger steamers in which the raising of ashes on deck is objectionable, the ashes are placed in a trough leading to a pipe, a jet of water at a pressure of about 200 pounds per square inch from one of the pumps is then admitted, and scours the ashes along the pipe into the sea. A small valve is fitted to permit the entry of air into the pipe during the discharge. The apparatus is simple and efficient.

pressure. In passing through only one door is open at a time which makes it possible to enter or leave the stoke-hold without allowing much air to escape and thus reduce the air-pressure in the stoke-hold.

Ques. 325.—At what places aboard a ship are air-locks necessary?

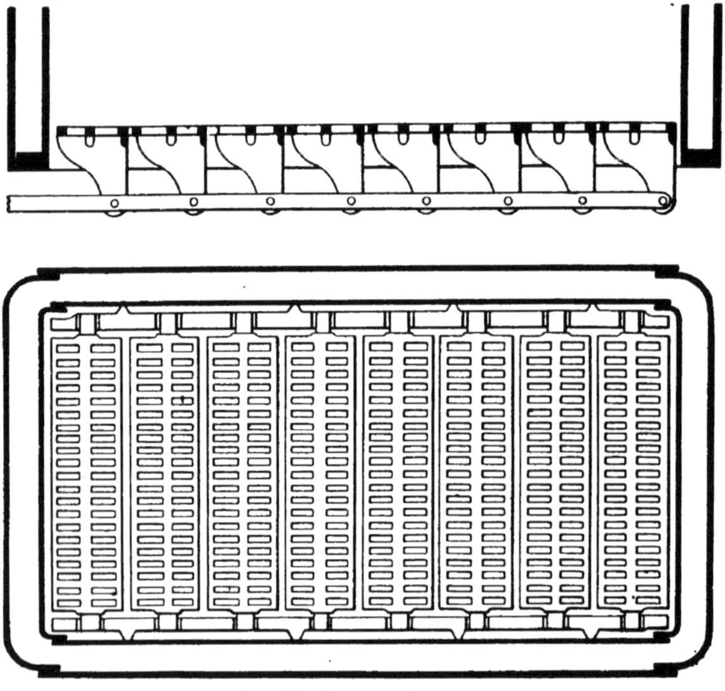

FIG. 89. SHAKING GRATES.

Ans.—At all places where communication is had between the compartments under pressure and any other part of the ship.

Ques. 326.—What are the advantages in general possessed by closed stoke-holds over other systems?

Ans.—First, a reduction in the space and weight

required by the boilers, since, by the addition of fans and screens, which are light and inexpensive, and supply the necessary air under pressure to the furnaces, the boilers may be made to develop from 20 to 25 per cent more power, than they would with natural draught; second, by the employment of blowing fans, a continuous supply of fresh air in the stoke-hold is assured and the health and comfort of the men working there is much better provided for than it would be with natural draught.

Ques. 327.—How are the ashes raised from the stoke-hold to the deck, to be thrown overboard?

Ans.—By means of the ash-tube and engine; the ash-tube leading from stoke-hold to deck, and the engine raising the ashes in an ash-bucket, that passes through the tube. Another method is by means of the ash-ejector, which is simply an inclined tube running from the stoke-hold to above the water-line, and overboard. At the lower end of this tube is a hopper, into which the ashes are shoveled, and at the bottom of this hopper they are picked up by a jet of water of high velocity, and forced through the inclined tube overboard.

CHAPTER V

BOILER OPERATION

Ques. 328.—What should be the first care of an engineer, or water-tender, when he goes on watch?

Ans.—He should ascertain the exact height of the water in his boilers by opening the valve in each of the drain-pipes of the water-columns, allowing it to blow out freely for a few seconds, then closing it tight, and allowing the water to settle back in the glass.

Ques. 329.—What is one of the important duties of the firemen coming off watch?

Ans.—They should have the fires clean, the ash-pits all cleaned out, a good supply of coal on the floor, and everything in good condition for the oncoming force.

Ques. 330.—What implements are needed for successfully and quickly cleaning a fire?

Ans.—A slice-bar, a fire-hook, a heavy iron or steel hoe, and a lighter hoe for cleaning the ash-pit.

Ques. 331.—How may these tools be made, so that they will be light and easy to handle and at the same time strong and durable?

Ans.—After the working ends have been fashioned to the desired shape, let each be welded to a bar of 1-inch or 1⅛-inch round iron 10 or 12 inches in length. Then take pieces of 1-inch or 1¼-inch iron pipe, cut to the length desired for the handles, and weld the shanks of the tools to one end of the pipe handles and to the other end weld a ring handle or a short cross-bar to facilitate handling the tools.

Ques. 332.—When a fire shows signs of being foul and choked, what should be done at once?

Ans.—Prepare to clean it by allowing one side to burn as low as possible, putting fresh coal on the other side alone.

Ques. 333.—Describe the process of cleaning a fire.

Ans.—When the first side has burned as low as it can, without danger of letting the steam-pressure drop too low, take the slice-bar and shove it in along the side of the furnace, on top of the clinker, and back to near the bridge-wall, then, using the door-jamb as a fulcrum, give it a quick, strong sweep across the fire, and the greater portion of the live coals will be pushed over to the other side. What remains of the coal not yet consumed can be pulled out upon the floor with the light hoe and shoveled to one side, to be thrown back into the furnace after the clinker is removed. Having thus disposed of the live coal, take the slice-bar and shove it in on top of the grates, under the clinker, loosening and breaking it up, after which take the heavy hoe and pull it all out upon the floor, where the intense heat contained in the clinker should be quenched by a helper, with a pail of water, or water discharged from a small rubber hose.

Ques. 334.—Having gotten one side of the fire cleaned, what is the next move?

Ans.—Close the door for that side, and with the slice bar in the other side, push all the live coal over to the side just cleaned, where it should be leveled off, and fresh coal added. After this has become ignited treat the other side in the same manner.

Ques. 335.—Can a definite code of rules for hand firing, be laid down, that will suit all conditions?

Ans.—No; owing to the fact that there are so many different varieties of coal, some of which need very little stirring or slicing, while others, that have a tendency to coke and form a crust on top of the fire, need to be sliced quite often.

Ques. 336.—Mention a few general maxims that are applicable to all boiler-rooms.

Ans.—First, keep a clean fire; second, see that every square inch of grate surface is covered with a good live fire; third, keep as level a fire as possible; fourth, when cleaning the fire, be sure to clear all the clinkers and dead ashes away from the back end of the grates at the bridge-wall.

Ques. 337.—Why should the face of the bridge-wall, especially, be kept clean and free from ashes and clinker?

Ans.—For the reason that this is one of the best points in the furnace for securing good combustion, provided that the bridge-wall is kept clean from the grates up, and by keeping the back ends of the grates clean, the air is allowed a free passage through them and is permitted to come directly in contact with the hot fire-brick, and thus one of the greatest aids to good combustion is utilized.

Ques. 338.—In firing bituminous coal, what is a good plan to pursue in regard to the fire-doors, with some kinds of boilers?

Ans.—They should be left slightly open for a few seconds, immediately after throwing in a fresh fire.

Ques. 339.—Give the reason for doing this.

Ans.—Bituminous coal contains a large percentage of volatile (light or gaseous) matter, which flashes into flame the instant it comes in contact with the live fire in the furnace, and if a sufficient supply of oxygen is not present just at this particular time, the combustion will be imperfect, and the result will be the formation of carbon monoxide, or carbonic oxide gas, and the loss of about two-thirds of the heat units contained in the coal.

Ques. 340.—How may this great loss of heat be guarded against, in a great measure?

Ans.—By admitting a sufficient volume of air, either through the fire-doors, directly after putting in a fresh fire, or what is still better, providing air-ducts through the bridge-wall, or side walls, which will bring the air in above the fire.

Ques. 341.—What quantity of air is required for the complete combustion of 1 pound of coal?

Ans.—By weight, 12 pounds; by volume, about 150 cubic feet.

Ques. 342.—Is there any advantage gained by heating this air before admitting it to the furnace?

Ans.—There is a great advantage, provided the heat used for this purpose would otherwise be wasted. Great economy in fuel, and much better combustion, result from supplying heated air to the furnaces.

Ques. 343.—Describe the Howden draught system, as used in the marine service.

Ans.—There is a nest of tubes in the uptake that is enveloped by the hot gases on their way to the stack. The air is caused to pass through these tubes by a

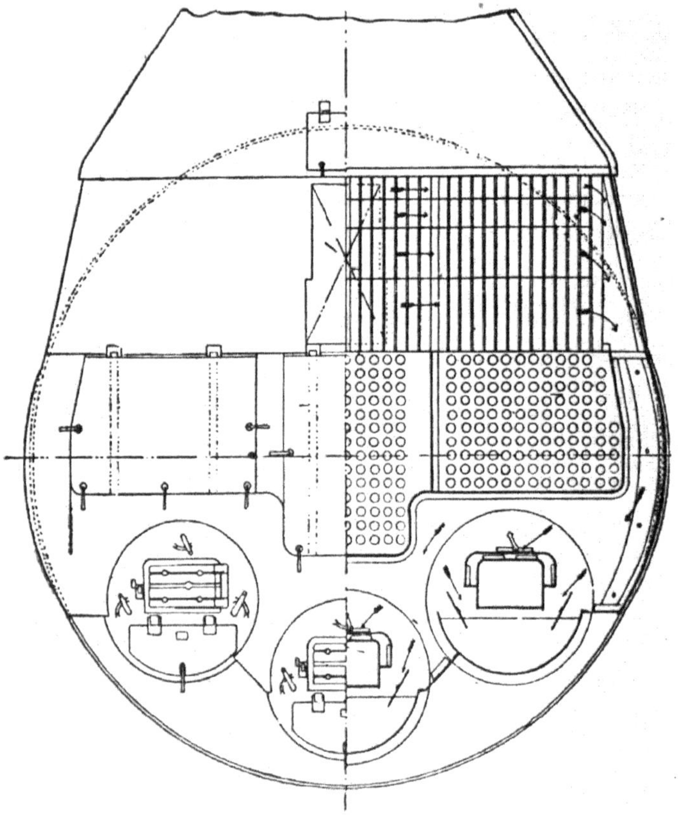

FIG. 90. ARRANGEMENT OF THE HOWDEN DRAUGHT SYSTEM.

blower-fan, and as a consequence is heated to a high degree before passing into the ash-pit. Some of this hot air is also directed into the furnace above the fire, thus securing a good combustion of the fuel.

Ques. 344.—What precautions should be taken regarding cleanliness of the tubes?

Ans.—The tubes of all boilers should be kept clean and free from soot, and especially does this apply to fire-

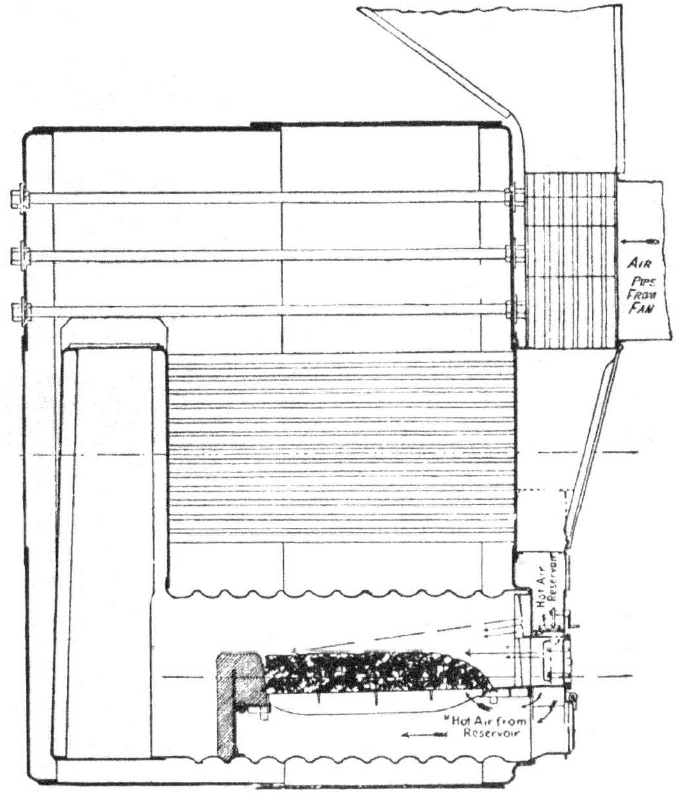

Fig. 91. Air Heater of the Howden Draught System.

tube boilers, for the reason that, when these tubes become clogged with soot, the efficiency of the draught is destroyed and the steaming capacity of the boiler is greatly reduced, because soot not only stops the draught but it is also a non-conductor of heat.

Ques. 345.—What methods are ordinarily employed for cleaning the soot and dust from tubes?

Ans.—First, the steam jet, if properly made and connected by steam hose so as to get dry steam of high pressure, will do very effective work; second, a scraper having steel blades expanded by springs so as to fit the inside of the tubes snugly, should be pushed through each tube once or twice during each twenty-four hours of service. This will cut the soot loose from the inside surface of the tubes, and greatly facilitate blowing it out with the steam jet. For the tubes of water-tube boilers

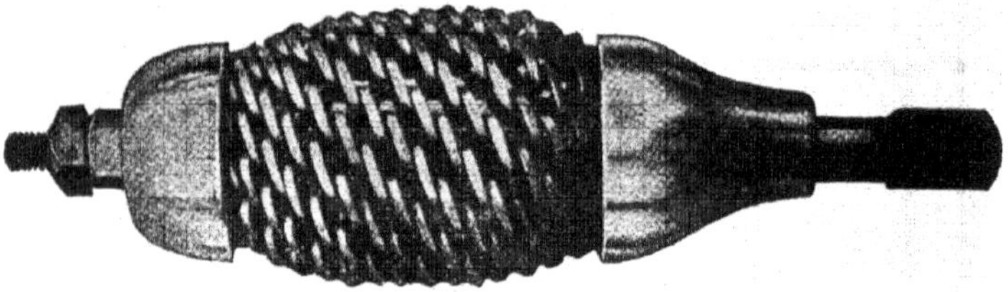

FIG. 92. SCRAPER FOR CLEANING FIRE TUBES.

the steam jet may be employed to advantage in cleaning the outside surfaces, and a rotary scraper driven by a small steam turbine is used for cleaning the scale formation from the inside.

Ques. 346.—How often should a boiler be washed out and cleaned inside?

Ans.—If the feed-water is impregnated to a considerable extent with scale-forming matter, the boiler should be washed out every two weeks, and if the water is very bad, the time should be shortened to one week.

Ques. 347.—How should a boiler be prepared for washing out?

Ans.—The fire should be allowed to burn as low as possible, and then be all pulled out of the furnace, the fire-doors left slightly ajar, and the dampers left wide open in order that the boiler may gradually cool.

Ques. 348.—Should a boiler be blown out, that is, emptied of water, while there is any steam-pressure in it?

Ans.—It should not.

Ques. 349.—Why not?

Ans.—For the reason that the sudden change of temperature from hot to cold has an injurious effect on the seams and braces. It is as bad a practice to cool a boiler down too suddenly as it is to fire it up too quickly.

FIG. 93. TURBINE CLEANER FOR WATER TUBES.

Ques. 350.—What effect does the too sudden contraction or expansion of the boiler-plates have upon the riveted seams?

Ans.—Leaks are created, and very often small cracks radiating from the rivet-holes are started, and these becoming larger with each change of temperature, will finally destroy the strength of the seam and serious results will follow.

Ques. 351.—Suppose that all of the fire has been pulled from the furnace and that the boiler has stood until the steam-gauge indicates 20 pounds pressure, would it then be safe to blow all of the water out of the boiler?

Ans.—It would not, for the reason that the temperature of steam at 20 pounds pressure is 260 degrees Fahrenheit, and it may be assumed that the temperature of the metal of the boiler is at or near this temperature also. Assuming the temperature of the atmosphere in the boiler-room to be 60 degrees Fahrenheit there will be a range of 260 degrees — 60 degrees = 200 degrees Fahrenheit temperature for the boiler to pass through within a short time, which will certainly have a bad effect, and besides this, the boiler shell will be so hot that the loose mud and sediment left after the water has run out is liable to become baked upon the bottom sheets, making it much harder to remove.

Ques. 352.—Under what conditions is it best to empty a boiler of water preparatory to washing it out?

Ans.—After the boiler has become comparatively cool and there is no pressure indicated by the steam-gauge, the blow-off cock may be opened and the water allowed to run out. The gauge-cocks and drip-valve to the water-column should be left open to allow the air to enter and displace the water, otherwise there will be a partial vacuum formed in the boiler, and the water will not run out freely.

Ques. 353.—Mention some of the important duties of the boiler-washer.

Ans.—After the water has all run out and the boiler has cooled sufficiently to permit it, he should go inside (provided there is a man-hole) and after having thoroughly cleaned the inside of the boiler, he should closely examine all of the braces and stays, and if any are found loose or broken, they should be repaired at once, before the boiler is put in service again. The soundness of braces, rivets, etc., can be ascertained by tapping them with a light hammer.

Ques. 354.—What should be done with the tubes of fire-tube boilers when they become coated with scale on their outside surfaces?

Ans.—The boiler should be taken out of service, laid up temporarily, and the tubes taken out, cleaned, and those that are not corroded or pitted too badly may be made almost as good as new by cutting off 8 or 10 inches of the ends and welding pieces of new tubing on, to bring the tubes back to their original length, after which they may be put back in the boiler and be good for a long term of service. While the tubes are out of the boiler for repairs the boiler-washer will have a good opportunity to get inside and clean and inspect every portion of the inside.

Ques. 355.—What precautions should be taken when connecting a recently fired-up boiler with the steam main or header?

Ans.—First, the steam in the boiler to be connected should be raised to the same pressure as that in the main, then the dampers should be closed and the steam stop-valve should be opened slightly, just enough to permit a small jet of steam to pass through, which can be heard by placing the ear near the body of the valve. This jet of steam may be passing from the main into the newly connected boiler, or vice versa. Whichever way it is

Fig. 94. Square Open Heater.

going, the valve ought not to be opened any farther until the flow of steam stops. This will indicate that the pressure has been equalized between the boiler and the main, and it will then be found that the valve will move much easier, and it may be gradually opened until it is wide open.

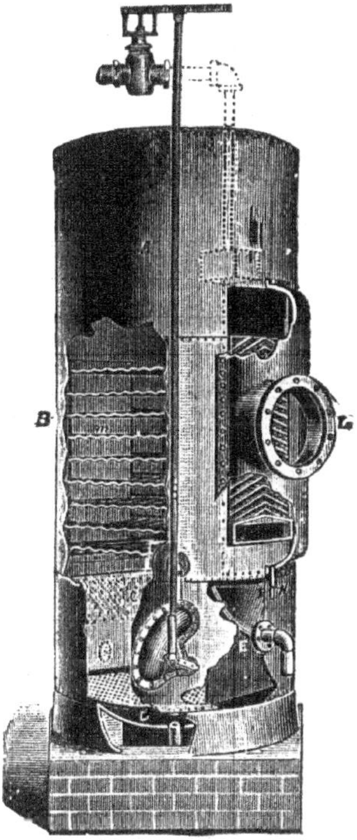

FIG. 95. INTERIOR VIEW OF OPEN HEATER.

Ques. 356.—Should cold feed-water ever be pumped into a boiler that is under steam?

Ans.—It should not, if it is possible to prevent it.

Ques. 357.—How may the feed-water be heated economically?

Ans.—By passing it through a feed-heater in which the heating agent employed is the exhaust steam from the engines.

Ques. 358.—How should the feed-water be supplied to a boiler while the boiler is being fired?

Ans.—It should be supplied just as fast as it is evaporated. The firing can then be even and regular.

Ques. 359.—If the supply of feed-water should suddenly be cut off owing to breakage of the pump or some other cause, and no other source of supply was available, what should be done?

Ans.—The dampers should be closed immediately, and all of the draught stopped. The fires should be deadened by shoveling wet or damp ashes in on top of them, or if ashes can not readily be procured, bank the fires over with green coal broken into fine bits. This, with the draught all shut off, will keep the fires dead, and if repairs to the feed-supply can not be made within a short time, the fires should be pulled, that is, if they have become deadened sufficiently.

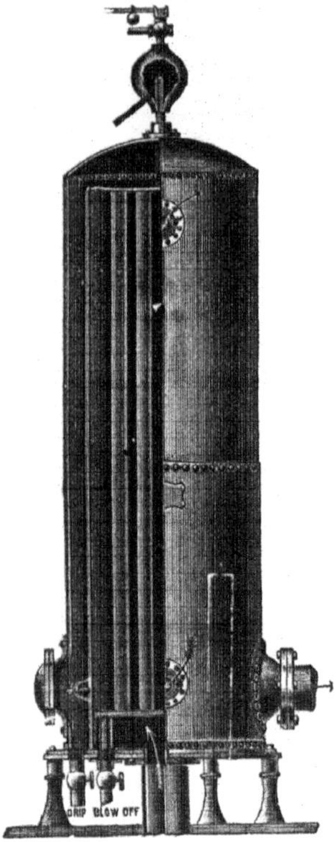

FIG. 96. BARAGWANATH STEAM JACKET FEED WATER HEATER.

Ques. 360.—Should the fires be pulled while they are burning lively?

Ans.—No; because the stirring will only serve to increase the heat, and the danger will be aggravated.

Ques. 361.—What is the primary object of making evaporation tests of boilers?

Ans.—To ascertain how many pounds of water per pound of coal the boiler is evaporating.

Ques. 362.—What other important details relating to the operation of the boilers may be ascertained through a well-conducted evaporation test?

Ans.—First, the efficiency of the boiler and furnace as

an apparatus for the consumption of fuel and the evaporation of water; second, the relative value of different varieties of coal, and other fuels, as heat-producers; third, whether the boilers, as they are operated under ordinary every-day conditions, are being operated as economically as they should be; fourth, in case the boilers, owing to an increased demand for steam, fail to supply a sufficient quantity, whether or not additional boilers are needed, or whether the trouble could be overcome by a change of conditions in the operation of the boilers.

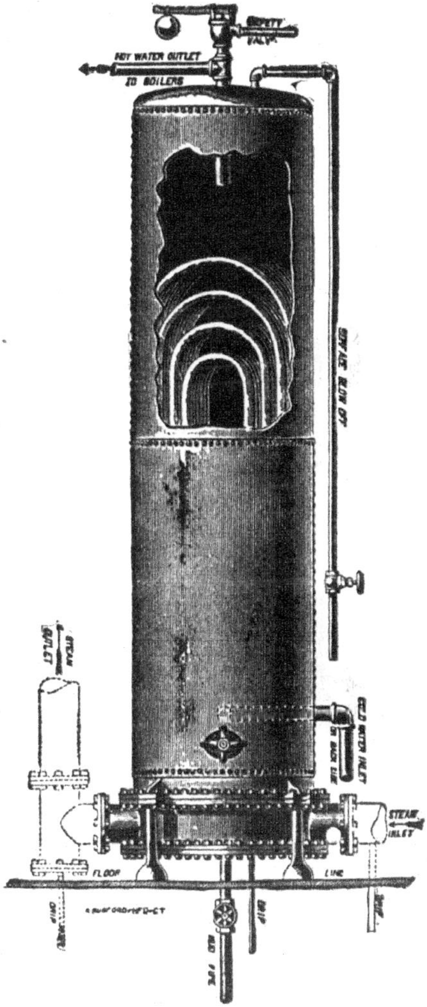

Fig 97. Closed Feed Water Heater.

Ques. 363.—What are the principal data to be noted down during the progress of an evaporation test?

Ans.—First, time—the number of hours that the test is conducted; second, the kind of coal burned; third, weight of coal consumed; fourth weight of water evaporated during the test; fifth, weight of dry ash returned; sixth, moisture in the coal per cent, seventh, dry coal corrected for moisture; eighth weight of combustible; ninth, moisture in the steam per cent; tenth, water corrected for moisture in the steam, eleventh, average temperature of the feed-water; twelfth, average temperature of the escaping gases; thirteenth, square feet of grate surface; fourteenth, square feet of heating surface; fifteenth, ratio of grate surface to heating surface.

Ques. 364.—How may the weight of the coal consumed during the test be ascertained?

Ans.—By having a small platform scales fitted with a wooden platform large enough to accommodate a wheel-barrow, or, in lieu of a barrow, a box large enough to contain two or three hundred pounds of coal. Each wheel-barrow load, or boxful of coal that goes to the boiler under test can then be weighed and the figures be placed upon a tally-sheet and added together at the close of the test, thus giving the total weight of coal consumed during the test. If, at the close of the test, there is any of the weighed coal left on the floor, it should be weighed back and deducted from the total weight.

Ques. 365.—How may the weight of water evaporated during the test be ascertained?

Ans.—By having a hot-water meter fitted in the branch feed-pipe leading to the boiler under test, or if this is not to be had, a substitute equally as accurate can be made by placing two small water-tanks, each having a capacity of 8 or 10 cubic feet, in the vicinity of the feed-pump.

These tanks can be made of light tank-iron, and each should be fitted with a nipple and valve, near the bottom, for connection with the suction side of the feed-pump. The tops of the tanks may be left open. A pipe leading from the main water-supply, with a branch to each tank, is also needed for filling them. If an open feed-water heater is used, and it is possible to place the tanks low enough to allow a portion of the hot water from the

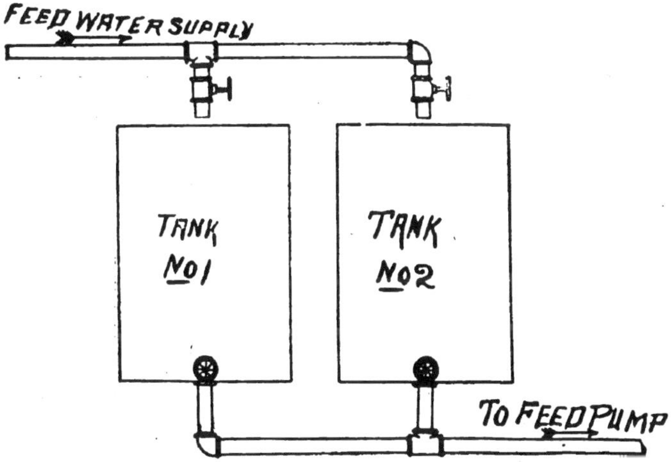

FIG. 98.

heater to be led into them by gravity, it will be desirable to do so. If this can not be done, some other provision should be made for at least partially warming the water before it goes to the boiler. The exact capacity of each one of these two tanks, either in cubic feet or in pounds of water, should be ascertained, and then all of the feed-water that is supplied to the boiler during the test is to be first passed through the tanks, which should be numbered one and two respectively, in order to prevent confusion in keeping a record of the number of tankfuls of water used during the test. Two tanks should be provided, in order that while the feed-pump is drawing the water from one, the other one may be filled. The feed-pump that is used to supply the boiler under test should have no connection whatever with the main feed-supply. By keeping tab of the number of tankfuls of water used during the test, and

TABLE 6

WEIGHT OF WATER AT VARIOUS TEMPERATURES

Temperature	Weight per Cubic Foot	Temperature	Weight per Cubic Foot	Temperature	Weight per Cubic Foot
32° F.	62.42 lbs.	132° F.	61.52 lbs.	230° F.	59.37 lbs.
42°	62.42	142°	61.34	240°	59.10
52°	62.40	152°	61.14	250°	58.85
62°	62.36	162°	60.94	260°	58.52
72°	62.30	172°	60.73	270°	58.21
82°	62.21	182°	60.50	300°	57.26
92°	62.11	192°	60.27	330°	56.24
102°	62.00	202°	60.02	360°	55.16
112°	61.86	212°	59.76	390°	54.03
122°	61.70	220°	59.64	420°	52.86

multiplying this by the capacity of each tank, the total weight of water evaporated is ascertained.

Ques. 366.—How is the weight of dry ash ascertained?

Ans.—No water should be allowed to come in contact with the ashes during the test, or if it is absolutely necessary to use water, it should be used as sparingly as possible, and as the ashes are pulled from the furnace or ash-pit, they should be thrown to one side, and allowed to become dry, after which the weight can be ascertained by means of the scales that was used for weighing the coal.

Ques. 367.—How is the amount of moisture in the coal ascertained?

Ans.—This can generally be obtained from the reports of the geologist of the state in which the coal was mined.

Ques. 368.—How is the weight of dry coal corrected for moisture ascertained?

Ans.—Deduct the percentage of moisture in the coal from the total weight of coal consumed.

Ques. 369.—How is the weight of combustible ascertained?

Ans.—Deduct the weight of dry ash returned from the weight of dry coal corrected for moisture.

Ques. 370.—How is the percentage of moisture in the steam determined?

Ans.—By means of an instrument called a calorimeter, or if such an instrument is not at hand, the condition of the steam as regards its dryness may be approximately estimated by observing its appearance as it issues from a pet-cock, or other small opening into the atmosphere. Dry, or nearly dry steam, containing about 1 per cent of moisture, will be transparent close to the orifice through which it issues, and if it is of a grayish white color it may be estimated to contain not over 2 per cent of moisture.

Ques. 371.—How is water corrected for moisture in the steam arrived at?

Ans.—Deduct the percentage of moisture in the steam from the total weight of water evaporated during the test.

Ques. 372.—How is the average temperature of the feed-water obtained?

Ans.—By means of a hot-water thermometer connected to the feed-pipe near to the check-valve, but between it and the feed-pump. If the thermometer is not attached to the feed-pipe, the temperature of the water in each tank should be taken and noted down, during the time that the feed-pump is drawing from it. From these notations, made at regular intervals during the progress of the test, the average temperature of the feed-water is easily calculated.

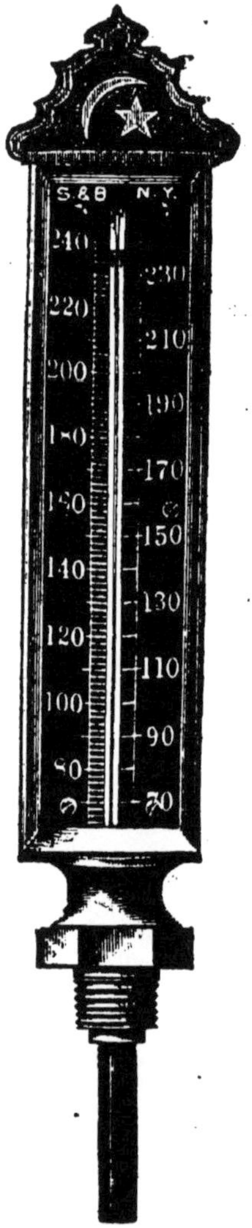

Fig. 99. Hot Water Thermometer.

Ques. 373.—How is the average temperature of the escaping gases determined?

Ans.—By readings taken at regular intervals from a thermometer connected in the uptake.

Ques. 374.—What should be done with the boiler and furnace before beginning an evaporative test?

Ans.—The boiler should be thoroughly cleaned, both inside and outside, and especially the heating surface, by scraping and blowing the soot out of the tubes, if it be a return-tubular boiler, and blowing the soot and ashes from between the tubes if it is a water-tube boiler. All dust, soot, and ashes should be removed from the outside of the shell, and also from the combustion chamber and smoke connections. The grate-bars and sides of the furnace should be cleared of all clinker, and all air-leaks made as tight as possible.

Ques. 375.—What should be done with the water-connections?

Ans.—The boiler and all of its water-connections should be perfectly free from leaks, especially the blow-off valve or cock. If any doubt exists as to the latter, it should be plugged, or a blind flange put on it.

Ques. 376.—Why is it required that especial care be exercised regarding the water-connections?

Ans.—For the reason that the test is made for the purpose of ascertaining the exact quantity of water that the boiler will evaporate with a given weight and kind of coal, and if any of the water fed to the boiler during the test is allowed to leak away, or if any water, other than that which has been measured by passing it through the tanks, is allowed to get into the boiler during the test, the results will be misleading and unreliable.

Ques. 377.—Before starting the test, what other details regarding the boiler should be attended to carefully?

Ans.—The boiler should be thoroughly heated, by having been run for several hours at the ordinary rate. The fire should then be cleaned and put in good condition to receive the fresh coal that has been weighed for the test.

Ques. 378.—What should be done regarding the water-level?

Ans.—At the time of beginning the test, the water-level in the boiler should be at or near the height ordinarily carried, and its position should be marked by tying a cord around one of the guard-rods of the gauge-glass, and, to prevent any possibility of error, the height of the water in the glass should be measured in inches, and a memorandum made of it.

Ques. 379.—What data regarding the steam-pressure should be recorded?

Ans.—The steam-pressure as indicated by the gauge should be noted at the time of starting the test, and also at regular intervals during the progress of the test, in order that the average pressure may be obtained.

Ques. 380.—When should the test begin?

Ans.—When all of the conditions just described have been complied with and the first lot of weighed coal has been fed to the furnace and the feed-pump is receiving water from one of the measuring tanks, the time should be noted and recorded as the starting time.

Ques. 381.—What length of time should an evaporation-test be conducted?

Ans.—Ten hours, if it is possible to continue it that long.

Ques. 382.—What conditions regarding the steam-pressure, condition of the fire and the water-level should prevail at the close of the test?

Ans.—They should be as nearly as possible the same at the close as they were at the beginning. The water-level should be the same and the quantity and the condition of the fire, also the steam pressure.

Ques. 383.—How may this be accomplished?

Ans.—Only by very careful work toward the close of the test.

Ques. 384.—If any of the weighed coal is left on the floor at the close of the test, what should be done with it?

Ans.—It should be weighed back and its weight deducted from the total weight.

Ques. 385.—If a portion of water is left in the last tank tallied, what disposition should be made of it?

Ans.—It should be measured and deducted from the total.

Ques. 386.—In making a test of the efficiency of the boiler, what is one of the most essential conditions to be taken into consideration?

Ans.—The boiler should be operated at its fullest capacity, from the beginning to the end of the test, and arrangements should be made to dispose of the steam as fast as it is generated.

Ques. 387.—How may this be done?

Ans.—If the boiler is in a battery and connected to a common header, the other boilers can be fired lighter during the test; but if there is but the one boiler in use, a waste-steam pipe should be temporarily connected, through which the surplus steam, if there is any, can be discharged into the open air, through a valve regulated as required.

Ques. 388.—If the boiler under test is fed by an injector instead of a pump during the test, from whence should the steam-supply for the injector be taken?

Ans.—The steam for the injector should be taken directly from the boiler under test, through a well-protected pipe. The steam for the pump, if one is used, should also be taken from the same source.

Ques. 389.—How should the temperature of the feed-water be taken when an injector is used?

Ans.—It should be taken from the measuring tanks, or at least from the suction side of the injector.

Ques. 390.—Why?

Ans.—Because the water in passing through the injector receives a large quantity of heat imparted to it by live steam directly from the boiler, and the temperature of the water after it leaves the injector would not be a true factor for use in calculating the results of the test.

Ques. 391.—For obtaining reliable and economical results in an evaporation-test, what conditions are essential regarding the draught?

Ans.—There should be a good, strong draught, which can be regulated by a damper, as desired. There should also be a draught-gauge connected to the uptake, for the purpose of measuring the draught.

Ques. 392.—Why is it necessary to measure the draught?

Ans.—The principal reason for measuring the draught is that in making comparative tests of the heating value of different varieties of coal, the conditions should be the same as near as possible in all of the tests made, and especially should this be the case with the draught. Therefore, by using a draught-gauge and measuring the draught during each test, there will be no uncertainty regarding this very important element.

Ques. 393.—Describe the construction and operation of a draught-gauge.

Ans.—The usual form of draught-gauge is a glass tube bent in the shape of the letter U. One leg is connected to the uptake by a small rubber hose, while the other leg is open to the atmosphere.

A scale marked in tenths of an inch is fitted between the two legs of the gauge. The glass tube is partly filled with water, which will, when there is no draught, stand at the same height in both legs, provided the instrument stands perpendicular, which is its normal position. When connected to the uptake, the suction caused by the draught will cause the water in the leg to which the hose is attached to rise, while the level of the water in the leg that is open to the atmosphere will be equally depressed, and the extent of the variation in fractions of an inch is the measure of the draught. Thus the draught is referred to as being .5 .7 or .75 inch.

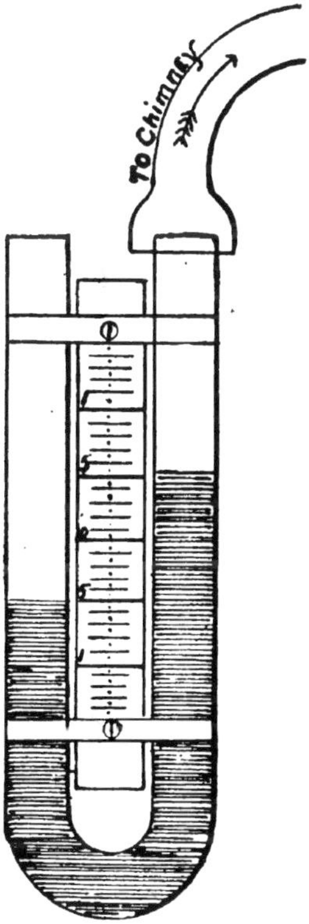

FIG. 100. DRAUGHT GAUGE.

Ques. 394.—What is the least draught that should be used, in order to obtain good results?

Ans.—The draught should not be less than .5 inch. Better results may be obtained with a draught of .7 inch.

Ques. 395.—If the test is made for the purpose of determining the efficiency of the boiler and setting as a whole, including grate, draught, etc., and also for comparing the heating qualities of different kinds of coal, what must the result be based upon?

Ans.—Upon the number of pounds of water evaporated per pound of coal burned.

Ques. 396.—What is implied in the expression "per pound of coal burned" as used in this connection?

Ans.—It includes not only the purely combustible matter in the coal, but the non-combustible also, such as ash, moisture, etc. Some varieties of Western coal contain as high as 12 to 14 per cent of moisture, and the ability of the furnace to extract heat from the mass is to be tested, as well as the ability of the boiler to absorb and transmit that heat to the water.

Ques. 397.—If the test is to determine the efficiency of the boiler itself as an absorber and transmitter of heat, what must be the factor for working out the result?

Ans.—The weight of the combustible alone must be considered.

Ques. 398.—When making a series of tests for the purpose of comparing the economical value of different kinds of coal, what conditions should prevail?

Ans.—The conditions should be as nearly uniform as possible; that is, let the tests all be made under ordinary working conditions, and with the same boiler or boilers, and if possible with the same fireman.

Ques. 399.—What is meant by the term "equivalent evaporation," as applied to the results of an evaporation-test?

Ans.—The term "equivalent evaporation," or the evaporation from and at 212 degrees, assumes that the feed-water enters the boiler at a temperature of 212 degrees and is evaporated into steam at 212 degrees temperature, and at atmospheric pressure, as, for instance, if the top man-hole plate were left out, or some other large opening in the steam-space of the boiler allowed the steam to escape into the atmosphere as fast as it was generated.

Ques. 400.—Why is it necessary to introduce this feature into calculations of the results of evaporation-tests?

Ans.—Owing to the variation in the average temperature of the feed-water used in different tests, and also the variation in the average steam-pressure, it is absolutely necessary that the results of all tests be brought by computation to the common basis of 212 degrees in order to obtain a fair and just comparison.

Ques. 401.—Describe the method of calculation by which this is done.

Ans.—Suppose an evaporation-test to have been made, and that the average steam-pressure by the gauge was 85 pounds, which equals 100 pounds absolute pressure, and that the average temperature of the feed-water was 141 degrees. By reference to Table 1, Chapter 1, it will be found that in a pound (weight) of steam at 100 pounds absolute pressure there are 1,181.1 heat units or thermal units, and in a pound of water at 141 degrees temperature there are 109.9 heat units. It therefore required 1,181.1 — 109.9 = 1,071.9 heat units to convert 1 pound of feed-water at 141 degrees temperature into steam at 85 pounds gauge, or 100 pounds absolute pressure. Now to convert a pound of water at 212 degrees temperature into steam at atmospheric pressure and 212 degrees temperature, requires (according to Table 1) 965.7 heat units, and the 1,071.9 heat units would evaporate 1,071.9 ÷ 965.7 = 1.11 pounds of water from and at 212 degrees. The 1.11 is the factor of evaporation for 85 pounds gauge pressure, and 141 degrees temperature of feed-water.

Ques. 402.—What use is made of this factor of evaporation in the calculation?

Ans.—One of the results of the test was "weight of water corrected for moisture in the steam," and by multiplying this result by the factor of evaporation, the "equivalent evaporation" is ascertained.

Ques. 403.—Upon what is the factor of evaporation based, in any test?

Ans.—Upon the steam-pressure and the temperature of the feed-water.

Ques. 404.—Give the formula for finding this factor for any test.

Ans.—The formula is: Factor $=\frac{H-h}{965.7}$, in which H = total heat in the steam, h = total heat in the feed-water, and 965.7 = the number of heat units in a pound of steam at atmospheric pressure and 212 degrees temperature. Table 7 gives the factor of evaporation, already calculated, for various pressures and temperatures.

Ques. 405.—If it is desired to ascertain the cost of coal for generating the steam used for operating an engine that uses 30 pounds of steam per horse-power per hour, what is the method of calculation?

Ans.—If the engine uses 30 pounds of steam per horse-power per hour, and it has been found by the test that 1 pound of the coal used would evaporate 9 pounds of water into steam of the pressure at which it is supplied to the engine, the actual consumption of fuel by the engine

TABLE 7

FACTORS OF EVAPORATION

Feed Water Temperature	Gauge Press. 50 lbs.	Gauge Press. 60 lbs.	Gauge Press. 70 lbs.	Gauge Press. 80 lbs.	Gauge Press. 90 lbs.	Gauge Press. 100 lbs.	Gauge Press. 110 lbs.	Gauge Press. 120 lbs.	Gauge Press. 140 lbs.
212°	1.027	1.030	1.032	1.035	1.037	1.039	1.041	1.043	1.047
200°	1.039	1.042	1.045	1.047	1.050	1.052	1.054	1.056	1.059
191°	1.049	1.052	1.054	1.057	1.059	1.061	1.063	1.065	1.069
182°	1.058	1.061	1.064	1.066	1.069	1.071	1.073	1.075	1.078
173°	1.067	1.070	1.073	1.076	1.078	1.080	1.082	1.084	1.087
164°	1.077	1.080	1.083	1.085	1.087	1.090	1.091	1.093	1.097
152°	1.089	1.092	1.095	1.098	1.100	1.102	1.104	1.106	1.109
143°	1.099	1.102	1.105	1.107	1.109	1.111	1.113	1.115	1.119
134°	1.108	1.111	1.114	1.116	1.119	1.121	1.123	1.125	1.128
125°	1.118	1.121	1.123	1.126	1.128	1.130	1.132	1.134	1.137
113°	1.130	1.133	1.136	1.138	1.140	1.143	1.145	1.146	1.150
104°	1.138	1.142	1.145	1.148	1.150	1.152	1.154	1.156	1.159
95°	1.149	1.152	1.154	1.157	1.159	1.161	1.163	1.165	1.169
86°	1.158	1.161	1.164	1.166	1.169	1.171	1.173	1.174	1.178
77°	1.167	1.170	1.173	1.176	1.178	1.180	1.182	1.184	1.187
65°	1.180	1.183	1.186	1.188	1.190	1.192	1.194	1.196	1.200
56°	1.189	1.192	1.195	1.197	1.200	1.202	1.204	1.206	1.209
47°	1.199	1.201	1.204	1.207	1.209	1.211	1.213	1.215	1.218
38°	1.208	1.211	1.214	1.216	1.218	1.220	1.222	1.224	1.228

would be as follows: $30 \div 9 = 3.33$ pounds of coal per horse-power per hour, which, multiplied by the total horse-power developed by the engine, will give the total weight of coal consumed in one hour's run.

Ques. 406.—What is the meaning of the expression "boiler horse-power?"

Ans.—The latest decision of the American Society of Mechanical Engineers regarding the horse-power of a boiler is "that the unit of commercial horse-power developed by a boiler shall be taken as 34½ units of evaporation." That is, 34½ pounds of water evaporated per hour from a feed temperature of 212 degrees into steam of the same temperature.

This standard is equivalent to 33,317 heat units per hour. It is also practically equivalent to an evaporation of 30 pounds of water from a feed temperature of 100 degrees Fahrenheit into steam of 70 pounds gauge-pressure.

Ques. 407.—According to this rule, what would be the horse-power of a boiler in which during a 10-hour test, the evaporation from and at 212 degrees was found by calculation to have been 86,250 pounds of water?

Ans.—The horse-power developed would be $86{,}250 \div 10 \div 34.5 = 250$ horse-power.

Ques. 408.—In what way can the maximum economy in the consumption of coal be obtained?

Ans.—There is only one way, and that is by keeping a continuous supply of coal on the fires and admitting a regular and sufficient quantity of air for its combustion.

Ques. 409.—Can these conditions be reached by hand firing?

Ans.—They can not, no matter how careful and skilful the firemen may be.

Ques. 410.—Mention two of the principal disadvantages attending hand firing.

Ans.—First, durin the time of firing the furnace

door is wide open, thus admitting a large volume of cold air; second, immediately after throwing in a fresh supply of coal, there is a sudden generation of gas, a large percentage of which escapes without being entirely consumed, and much heat is thus wasted.

Ques. 411.—What are the principles governing the operation of mechanical or automatic stokers?

Ans.—First, a continuous supply of coal and air; second, thorough regulation of the supply of fuel and air, according to the demand upon the boilers for steam; third, the intermittent opening and closing of the furnace doors is entirely prevented.

Ques. 412.—What are some of the disadvantages attending the use of mechanical stokers?

Ans.—First, the great cost of installing them; second, in case of a sudden demand upon the boilers for more steam, the mechanical stoker can not respond as promptly as in hand firing; third, the extra cost for power to operate them.

Ques. 413.—How many different classes of mechanical stokers are in use?

Ans.—Four general classes.

Ques. 414.—Describe the construction and operation of stokers belonging to Class 1.

Ans.—The grate consists of an endless chain of short bars, that travels in a horizontal direction from the front to the back of the furnace, over sprocket wheels operated either by a small auxiliary engine or by power derived from an overhead line of shafting in front of the boilers. The motion of the endless chain of grates is of course very

FIG. 101. BATTERY OF BABCOCK AND WILCOX WATER-TUBE BOILERS FITTED WITH CHAIN GRATE STOKERS.

slow, but it is continuous and regular, receiving the supply of coal at the front and depositing the ashes at the back end, where they drop into the ash-pit.

FIG. 102. MANSFIELD CHAIN GRATE STOKER, SHOWING HOW IT CAN BE WITHDRAWN FROM UNDER BOILER.

Ques. 415.—What type of stokers is included in Class 2?

Ans.—Stokers having grate-bars somewhat after the ordinary hand-fired type, but having a continuous motion up and down, or forward and back. Although this motion is slight, it serves to keep the fuel stirred and loosened, thus preventing the fire from becoming sluggish.

Ques. 416.—What position do the grate-bars in Class 2 occupy?

FIG. 103. VICARS MECHANICAL STOKER.

Ans.—Either horizontal, or slightly inclined, and their constant motion tends to gradually advance the coal from the front to the back end of the furnace.

Ques. 417.—What kinds of stokers are included in Class 3?

Ans.—Stokers in which the grates are steeply inclined.

The coal is fed onto the upper ends of the gates, which, having a slow motion, gradually force the coal forward as fast as required. In some stokers of this class, as, for instance, the Murphy, the grates incline from the sides towards the middle of the furnace, but in the majority of cases the inclination is from the front towards the back.

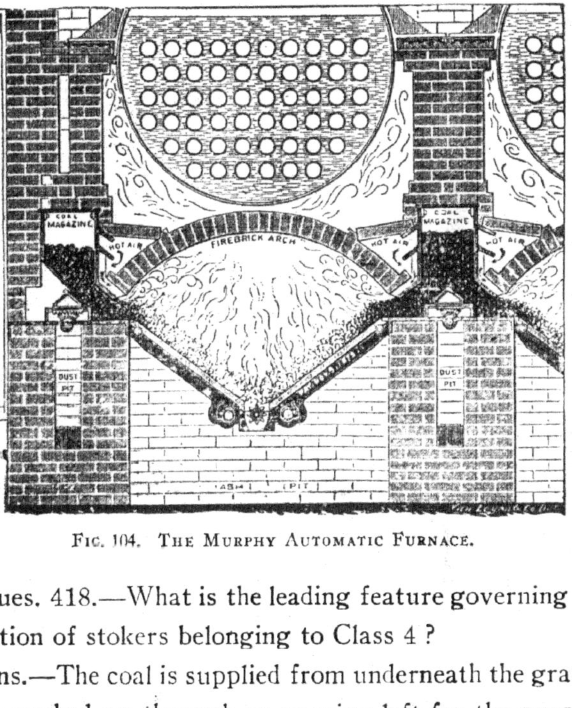

FIG. 104. THE MURPHY AUTOMATIC FURNACE.

Ques. 418.—What is the leading feature governing the operation of stokers belonging to Class 4 ?

Ans.—The coal is supplied from underneath the grates, and is pushed up through an opening left for the purpose midway of the length of the furnace. The gases, on being distilled, come in contact immediately with the hot bed of coke on top, and the result is good combustion.

Ques. 419.—What are stokers belonging to Class 4 called?

Ans.—Under-feed stokers.

Ques. 420.—What methods are employed for forcing the coal up into the furnace with under-feed stokers?

Fig. 105. Typical Setting of American Stoker Under a Return Tubular Boiler.

Ans.—Steam is the active agent, either by means of a steam-ram, or a long, slowly revolving screw, driven by a small engine.

Ques. 421.—How is the air supplied when an under-feed stoker is used?

Fig. 106. Standard 8 and 9 inch American Stokers.

A. Coal hopper.
B. Conveyor pipe.
C. Tuyeres for introducing air to the fuel.
D. Opening to wind-box for air connections.
E. Wind-box for supplying air to tuyeres.
K. Automatic steam-motor for driving conveyor.
R. Air-pipe to wind-box.
S. Gas-ducts for returning volatile products from entering coal to furnace.

Ans.—Forced draught is employed, and the air is blown into the furnace through tuyeres.

Ques. 422.—How is the coal supplied to mechanical stokers, other than the under-feed type?

Ans.—In two ways; either by being shoveled by hand

into hoppers in front of and above the grates, or, as is the case in most of the large plants using them, it is elevated by machinery and deposited in chutes, through which it is fed to each boiler by gravity. The coal used in mechanical stokers is in the form of screenings or nut coal.

Ques. 423.—Have mechanical stokers for feeding coal been applied in the marine service to any great extent?

Ans.—They have not, up to the present time.

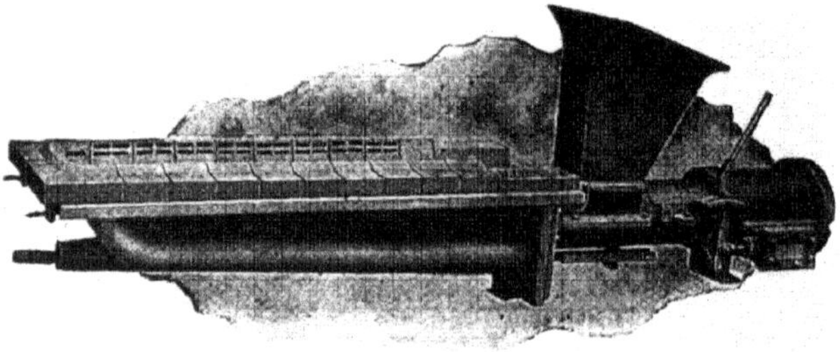

FIG. 107. JONES UNDER-FEED STOKER, HAVING A STEAM RAM.

Ques. 424.—In what way is it possible to successfully use automatic or mechanical stokers on marine boilers?

Ans.—By the use of liquid fuels, such as petroleum, blast-furnace oil, tar oil, etc.

Ques. 425.—Of what does petroleum consist?

Ans.—Petroleum consists practically of carbon, hydrogen, and oxygen, in the following proportions: Carbon, 85 per cent; hydrogen, 13 per cent, and oxygen, 2 per cent.

Ques. 426.—What is the heating value of 1 pound of petroleum?

Ans.—About 20,000 heat units, or about one-third more than the best coal.

Ques. 427.—How is petroleum fed to the furnaces?

Ans.—By being forced through nozzles having two or three holes, or annular spaces, from one of which the petroleum flows out, under pressure, while a jet of steam or compressed air issuing from another orifice catches the oil and "pulverizes" it into a fine spray, in which form it strikes the fire. The air for combustion is admitted through a third orifice, or if not thus supplied,

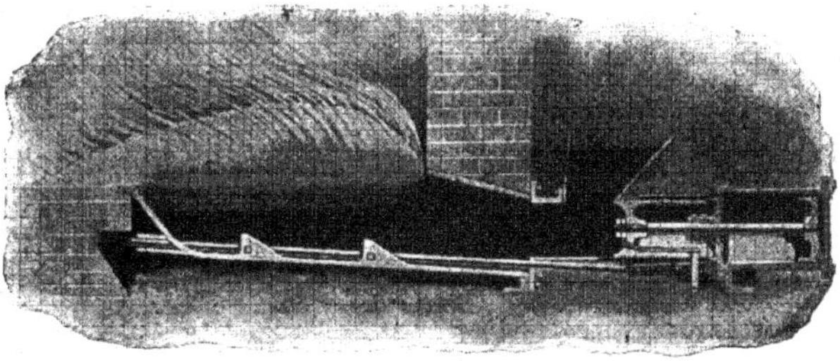

FIG. 108. SECTIONAL VIEW OF JONES UNDER-FEED STOKER.

the air for combustion is admitted by suitable orifices in the furnace front.

Ques. 428.—How is the furnace arranged for burning petroleum?

Ans.—A layer of broken fire-brick or asbestos is placed on the grate, and fire-brick screens, or bafflers, are placed in the way of the flame, thus providing a red-hot surface against which it impinges. Otherwise the combustion would be greatly hindered by the comparatively cool surfaces of the boiler-plates and tubes.

Ques. 429.—What agent has been found to be the best

for pulverizing the petroleum and spraying it into the furnace?

Ans.—Compressed air, slightly heated.

Ques. 430—What is one of the disadvantages attending the use of steam for this purpose?

Ans.—The danger of the flame being extinguished by water that is sometimes carried over with the steam.

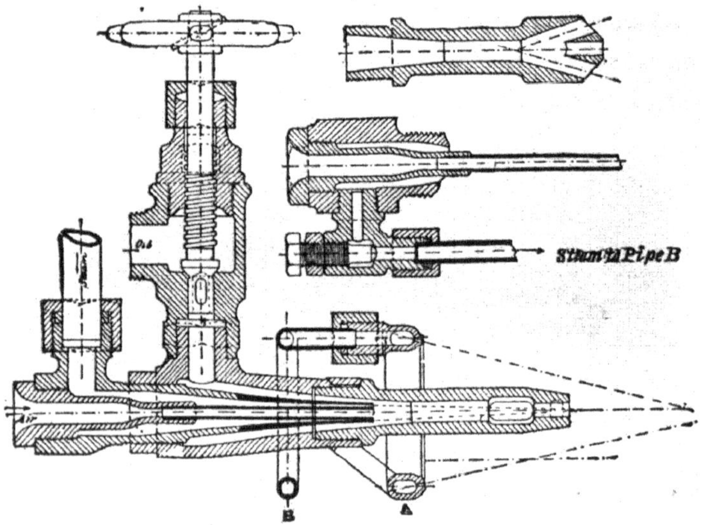

FIG. 109. PETROLEUM BURNER FOR BOILER FURNACE.

Ques. 431.—How is the oil supplied to the nozzles?

Ans.—By means of pumps that draw it from the bunkers and discharge it into a reservoir, and from thence it is fed to the burners.

Ques. 432.—What are the advantages in favor of petroleum fuel, especially for the marine service?

Ans.—First, superior evaporation, and, as a consequence, great reduction in the weight of fuel to be carried;

second, less space occupied by the fuel and ease of shipping it into the bunkers; third, reduction of stoke-hold force, also less space required in the stoke-hold; fourth, regularity of combustion and no reduction of power, due to cleaning fires, they being always clean and in a good condition; fifth, increased durability of boilers, owing to the fact that there are no variations of temperature, due to opening fire-doors, for coaling or cleaning; sixth, greater control over the expenditure of fuel, consequently less waste of steam at the safety valves, also less danger of a short supply of steam in case of a sudden demand upon the boilers.

Ques. 433.—What are some of the principal objections to its use on board of vessels?

Ans.—First, limited supply; second, vessels proceeding on long voyages could not, with present facilities, replenish their bunkers when required; third, danger of the generation of inflammable gases; fourth in war-ships the risk of possible loss of the fuel, in the event of injury to the bunker containing it.

Ques. 434.—Has the combination of coal and petroleum, in the same furnace, ever been attempted?

Ans.—Experiments along this line are being made in the British and other navies.

Ques. 435.—How are the furnaces fitted for this purpose?

Ans.—The same as for hand firing with coal, and in addition, a number of nozzles are placed in the front above the fire for injecting the petroleum over the incandescent coal.

Ques. 436.—Upon what does the efficiency of a steamship, or of a manufacturing establishment in which steam is used for power, largely depend?

Ans.—Upon the condition of the boilers and the care and labor expended for their preservation.

Ques. 437.—What was formerly one of the most dangerous and active agents in the deterioration of boilers, especially in the marine service?

Ans.—Corrosion of the boiler-plates and stays.

Ques. 438.—What was found, by a long series of experiments, to be the principal cause of this corrosion?

Ans.—The action of the fatty acids evolved by saponification from the heated tallow and vegetable oils, used at the time for the internal lubrication of the cylinders and valve-chests.

Ques. 439.—How were these oils carried into the boilers?

Ans.—In condensing systems, where the water of condensation was used for feed-water, the waste oil in the exhaust steam mingled with the feed-water and was carried into the boilers.

Ques. 440.—How has the danger from this source been largely obviated in late years?

Ans.—By the use of mineral oils for internal lubrication.

Ques. 441.—In what other way has the danger of corrosion been lessened?

Ans.—By the use of mild steel instead of iron plates in the construction of boilers. This steel is made by the

Seimens-Martin process, and is much stronger than iron and less liable to corrosive action.

Ques. 442.—Of what material are marine boilers now made entirely?

Ans.—Of steel, except the tubes, which are usually made of iron in the mercantile service. Steel tubes are used in war-ships. The furnaces and internal parts, that have to be welded or flanged, are made from specially soft steel plates.

Ques. 443.—What is the principal cause of corrosion in boilers at the present time?

Ans.—Oxidation of the plates, which results from contact with moisture and air, either carried in with the feed-water, or existing in the atmosphere when the boilers are empty.

Ques. 444.—What conditions must exist in order that corrosion shall take place from this cause?

Ans.—The simultaneous presence of both air and water, because neither dry air nor fresh water in which there is absolutely no air has any chemical action on steel or iron. Air dissolved in water is especially active, and the action is increased by the presence of various chlorides, such as magnesium and sodium.

Ques. 445.—Are there any other causes that tend toward the corrosion of boilers, internally?

Ans.—There are; for instance, hot sea-water, even when entirely deprived of air, has some action on steel and iron. It has been found that at the high temperatures now common in boilers, the chloride of magnesium contained in sea-water is decomposed by the heat and gives off hydrochloric acid, the evolution of acid being accelerated with increase of density.

Ques. 446.—Should sea-water be admitted to boilers if it is possible to prevent it?

Ans.—It should not; but if a portion is used, it is important that sufficient alkali, preferably lime, be admitted with the feed-water to render the water in the boilers slightly alkaline by the litmus test.

Ques. 447.—Does galvanic action, due to differences in the material used in the construction of the boilers, conduce towards corrosion?

Ans.—Galvanic action is probably a minor cause of corrosion.

Ques. 448.—What are some of the methods that may be employed for the prevention of corrosion in boilers?

Ans.—First, the admittance of air into the boilers while at work should be prevented as much as possible. This may be done by having a tank, called the feed-tank, into which the air-pumps may discharge its water, and from which the feed-pumps can draw their water for supplying the boilers. The feed-pumps should be independent pumps, which can be so regulated in speed as to be always fully supplied with water and never to empty the feed-tank and so suck in and discharge air into the boilers. Second, the complete exclusion of sea-water from the boilers if possible. The waste of feed-water should be made good by evaporators and a reserve of fresh water in tanks. Third, mineral oils, which consist of hydrocarbons only, should be used exclusively for lubrication of all internal parts of the engines and pumps requiring lubrication.

Ques. 449.—How may the injurious effects of galvanic action be neutralized?

Ans.—By the suspension of zinc slabs in various parts of the boiler, both below the water-line and also in the steam-space. Then if there be any galvanic action the zinc slabs will be attacked instead of the material of the boiler itself.

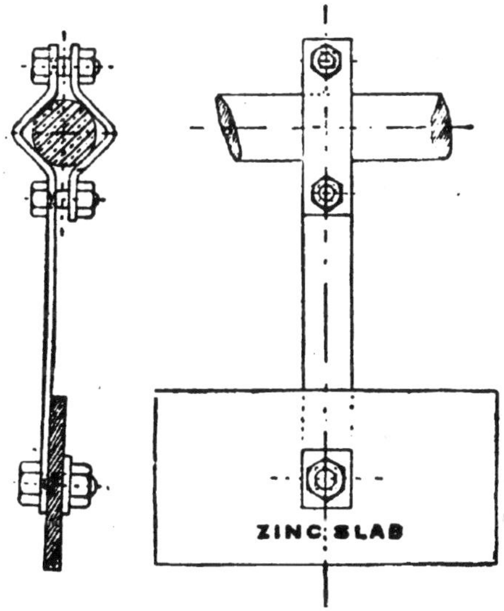

FIG. 110. METHOD OF SUSPENDING ZINC SLABS.

Ques. 450.—What is an important point to be observed when placing these zinc slabs?

Ans.—They should be in actual bright contact with the material of the boiler and they should be well distributed, so that every portion of the interior surface of the boiler is protected.

Ques. 451.—What is the theory of the action of these zinc slabs, in preventing galvanic corrosion?

Ans.—Zinc is an electro-positive metal, and it being suspended in the boiler causes the steel of the boiler to become electro-negative and thus any corrosive agent is induced to attack the zinc, leaving the steel uninjured.

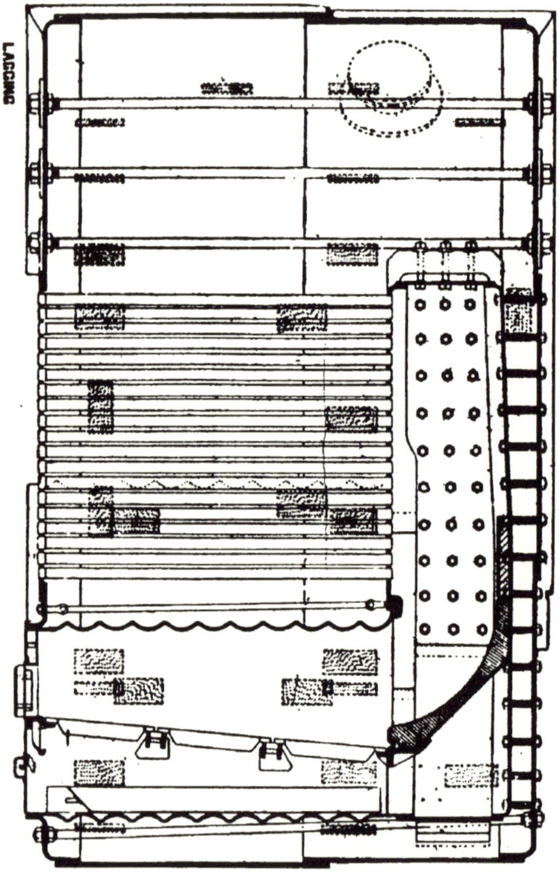

FIG. 111. SECTION OF MARINE BOILER, SHOWING LOCATION OF ZINC SLABS.

This preservative action can only take place when the boilers have water in them and the zinc slabs fitted in the steam-space act only when the boilers are completely filled with water.

CHAPTER VI

TYPES OF ENGINES—CLASSIFICATION

Ques. 452.—Into what three general classes may marine and stationary engines be divided?

Ans.—First, simple; second, compound; third, triple, or quadruple expansion.

Ques. 453.—What causes the piston of a steam-engine to move back and forth in the cylinder?

FIG. 112. CROSS COMPOUND DIRECT CONNECTED CORLISS ENGINE, ALLIS CHALMERS CO.

Ans.—The expansive force of the steam that is admitted alternately behind the piston, at either end of the cylinder.

Ques. 454.—Describe the action of the steam in a simple engine.

Ans.—In a simple engine the steam is used in but one cylinder, and from thence it is exhausted, either into the atmosphere or into a condensor.

Ques. 455.—What is the leading characteristic of a compound engine?

Ans.—In a compound engine the steam is made to do work in two or more cylinders before it is allowed to exhaust.

Ques. 456.—How is this accomplished?

Ans.—The compound engine is fitted with two, and in some cases with three cylinders. The cylinder into which steam at boiler pressure is admitted is termed the high-pressure cylinder and is the smallest of the group, in

FIG. 113. TANDEM COMPOUND ENGINE, BUCKEYE ENGINE CO.

diameter. The exhaust passage (or receiver) from this cylinder leads directly to the valve-chest of another cylinder, larger in diameter, termed the low-pressure cylinder, and thus conducts the exhaust from the high to the low-pressure cylinder, wherein it again serves as working steam, and if the cylinders are properly proportioned for the pressure, the amount of work done in each cylinder will be the same.

Ques. 457.—How many kinds of compound engines are in use generally?

Ans.—Two kinds: First, the cross compound, in which the cylinders stand parallel, each having its individual cross-head, connecting rod, and valve-gear, and all connected to a common crank-shaft; second, tandem

Fig. 114. Reynolds Combined Vertical and Horizontal Engine, 12,000 Horse-Power Cylinders, 4x88x60.

compound, in which the cylinders are tandem to each other, and one piston rod, cross-head, connecting rod, and valve-gear is common to both, although each cylinder has its own valve or valves for controlling the admission and release of the steam.

Ques. 458.—What is implied in the expression "triple expansion?"

Ans.—Triple expansion means that the steam has been allowed to expand through three successive stages, doing

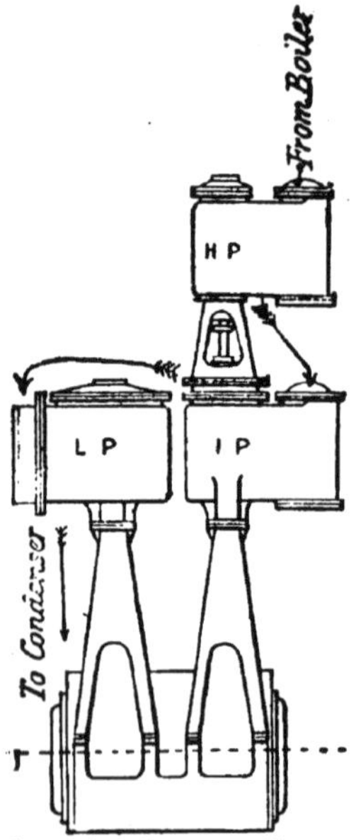

FIG. 115. SHOWS A TRIPLE EXPANSION ENGINE IN WHICH THE HIGH PRESSURE IS TANDEM WITH THE INTERMEDIATE CYLINDER.

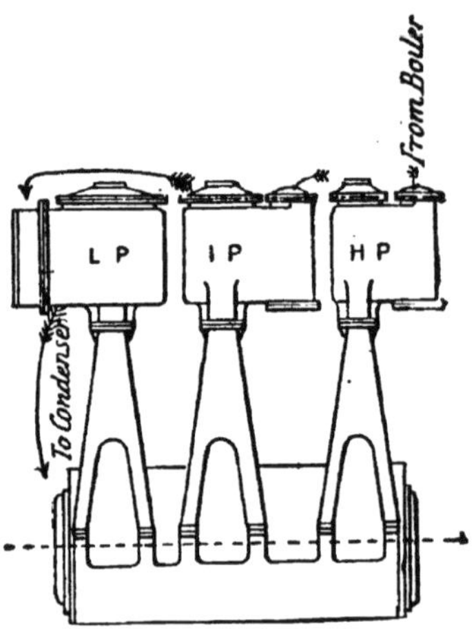

FIG. 115A. SHOWS THE ORDINARY ARRANGEMENT OF CYLINDERS FOR A TRIPLE EXPANSION ENGINE.

a fixed amount of work in each stage, before release occurs.

Ques. 459.—How many cylinders are required on a triple-expansion engine?

Ans.—Never less than three, and for large, high-speed engines it often becomes necessary to have two low-pressure cylinders, thus making a four-cylinder triple-expansion engine.

Ques. 460.—Are four cylinder triple-expansion engines much in use?

Ans.—They are in the marine service, and especially in the British navy, and they are also used to a large extent in the mercantile service.

Ques. 461.—Describe the action of the steam in a quadruple-expansion engine.

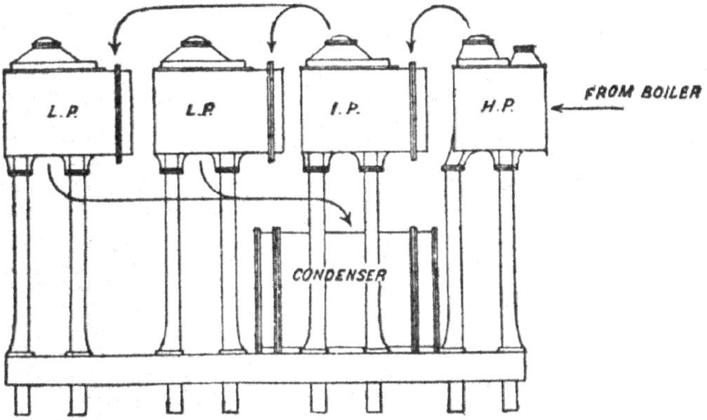

FIG. 116. ARRANGEMENT OF FOUR CYLINDER TRIPLE EXPANSION ENGINE FOR MARINE SERVICE.

Ans.—In a quadruple-expansion engine, the expansion of the steam is divided up into four stages by causing it to pass through four successive cylinders, termed respectively the high-pressure, first intermediate, second intermediate, and low-pressure. In some of the larger engines of this type there are two low-pressure cylinders, thus making five cylinders in all.

Ques. 462.—What pressures of steam are usually used in this type of engine?

Ans.—From 200 to 250 pounds per square inch.

Ques. 463.—What are some of the advantages that are to be gained in the use of steam by stage expansion?

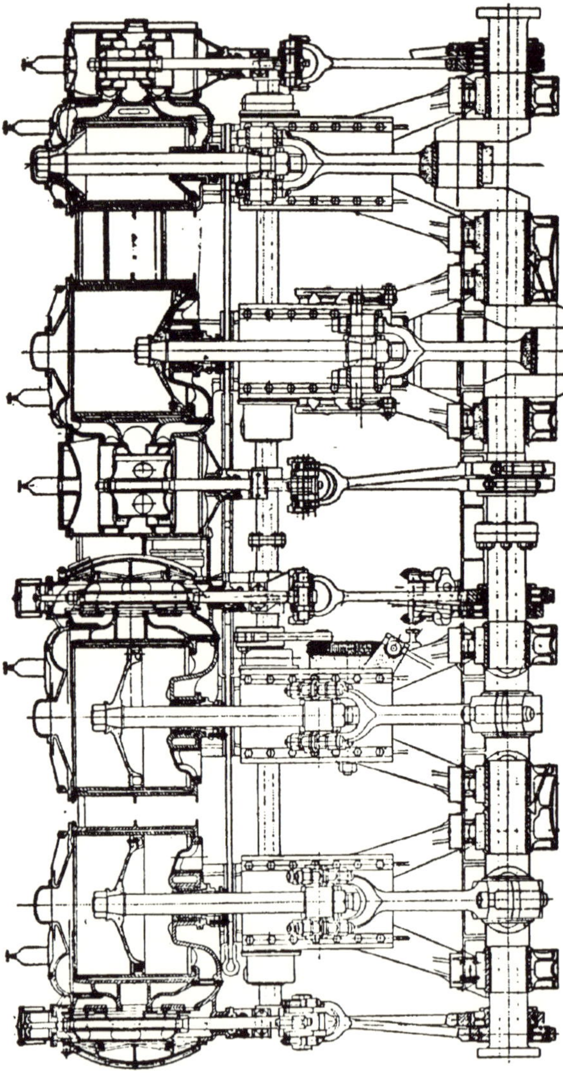

Fig. 117. Sectional View of Four Cylinder Triple Expansion Marine Engine for Large Ocean-Going Steamer.

Ans.—First, that the cylinder into which steam directly from the boiler is admitted is never open to the cooling influence of the atmosphere, or condensor, hence there is not so much cooling and condensation of the entering steam; second, the steam that is condensed and reevaporated in the first cylinder reappears as working steam in the second cylinder; third, the loss from condensation in the second and third cylinders is also reduced, owing to the smaller range of temperature, between admission and exhaust in those cylinders.

Ques. 464.—What are the mechanical advantages of compound and triple-expansion engines, for heavy duty?

Ans.—First, the facility with which high rates of expansion may be carried out without bringing excessive strains and stresses on the framing of the engine; second, a greater uniformity of twisting moment on the shaft.

Ques. 465.—What are the usual ratios of cylinder volumes in compound and triple and quadruple-expansion engines?

Ans.—For compound engines 1 to 4 between high and low-pressure cylinders. For triple-expansion engines, the ratios are about 1, 3 and 7, for high, intermediate and low-pressure cylinders. For quadruple-expansion engines the ratios are as follows: 1, 2, 4½ and 10¼ for high-pressure, first intermediate, second intermediate and low-pressure respectively.

Ques. 466.—What is meant by the term receiver, as used in connection with the stage-expansion of steam?

Ans.—In the case of a compound engine the receiver is the whole of the space between the high-pressure piston, when at the end of its stroke, and the back of the low-pressure steam-valve, whether it be slide rotative, or piston-valve. In the case of a triple-expansion engine, the space between the piston at the end of its stroke and the back of the intermediate steam-valve is called the intermediate receiver, and the space between the inter-

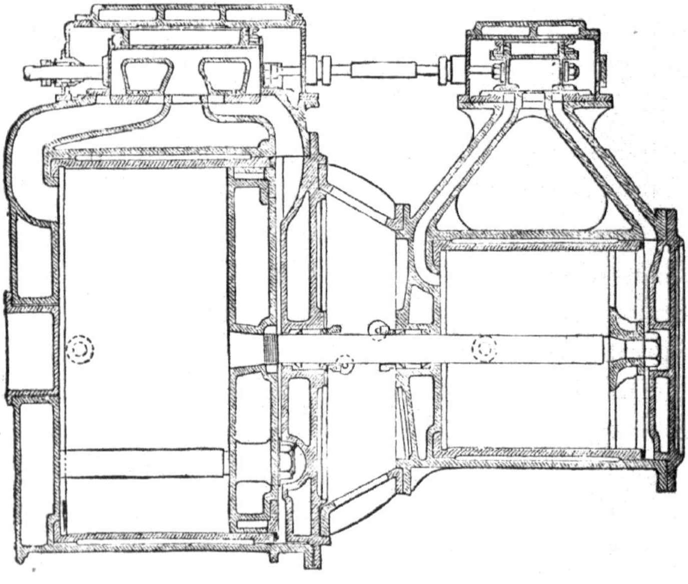

FIG. 118. SECTIONAL VIEW OF TANDEM COMPOUND CYLINDERS, SHOWING ARRANGEMENT OF STEAM CHESTS AND VALVES.

mediate piston at the end of its stroke and the low-pressure steam-valve is the low-pressure receiver.

Ques. 467.—What is the usual volume of these receivers in modern practice?

Ans.—After many experiments with large reservoirs as receivers, it has been found that all that is necessary is a comparatively large exhaust pipe from the exhaust orifice of the high-pressure cylinder to the steam inlet of the next lower pressure cylinder, it having been demonstrated that the volume of the exhaust passage and pipe from the high-pressure cylinder and the low-pressure valve-chest supplied sufficient space to allow for the compression that occurs between release from the high-pressure cylinder and admission to the low-pressure cylinder.

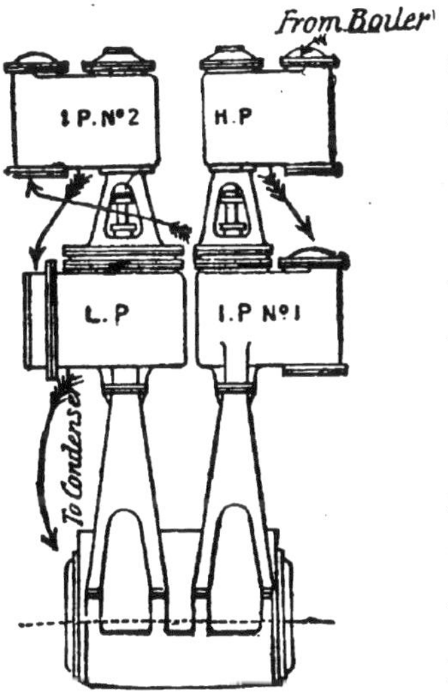

FIG. 119. TANDEM QUADRUPLE EXPANSION MARINE ENGINE SHOWING ARRANGEMENT OF CYLINDERS.

Ques. 468.—Does this law apply in the case of triple and quadruple-expansion engines?

Ans.—It does.

Ques. 469.—Upon what does the power of any stage-expansion engine depend?

Ans.—The power of a stage-expansion engine working at any given rate of expansion depends entirely upon the dimension of its low-pressure cylinder or cylinders, and is not affected by the size of its high-pressure cylinder, which latter, in fact, carries out but one stage in the expansion.

Ques. 470.—What does the capacity of the low-pressure cylinder or cylinders of such an engine require to be?

Ans.—The same as that of the whole of the cylinders of a simple engine of the same power, working at the same initial pressure and total ratio of expansion.

Ques. 471.—Why is this?

Ans.—For the reason that, since the initial pressures and ratios of expansion are the same in both engines, it follows that the terminal pressures and volumes must also be identical in both cases. In the simple engine the whole of the steam at the end of the stroke fills all of the cylinders, while in the compound engine it is contained in the low-pressure cylinder or cylinders only, hence the capacity of this cylinder must be equal to the capacity of all the cylinders of the simple engine.

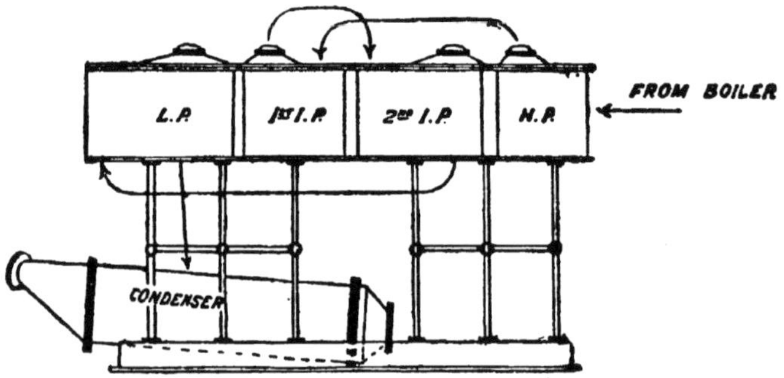

FIG. 119A. QUADRUPLE EXPANSION ENGINE, WITH CYLINDERS AS ORDINARILY ARRANGED—ARROWS SHOW COURSE TAKEN BY THE STEAM.

Ques. 472.—Why is it necessary in some cases to employ two low-pressure cylinders?

Ans.—For the reason that in very large engines one low-pressure cylinder would be too large and unwieldy, therefore it is divided into two equal parts.

Ques. 473.—Are compound and triple-expansion engines much in use outside of the marine service?

Ans.—They are to a large extent, owing to the great gain in economy over the simple engine. Practically all large manufacturing plants use them.

Ques. 474.—What other types of engines are in use in the marine service?

Ans.—The vertical walking-beam engine is largely in use on the lakes, bays, and rivers of the United States.

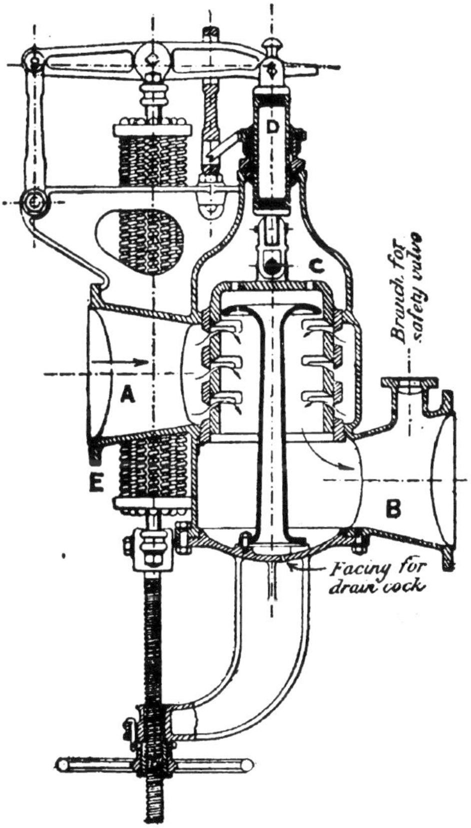

FIG. 120. BELLEVILLE REDUCING VALVE.

Ques. 475.—What is the leading characteristic of this type of engine?

Ans.—It has usually but a single cylinder, with a very

long stroke in proportion to its diameter, the length of the stroke varying from 7 to 12 feet.

Ques. 476.—What pressures of steam are usually employed in beam engines?

Ans.—Owing to the fact that the steam is expanded in a single cylinder only, the pressure carried is low—50 to 60 pounds per square inch.

Ques. 477.—Mention another type of engine that is in common use on Western rivers.

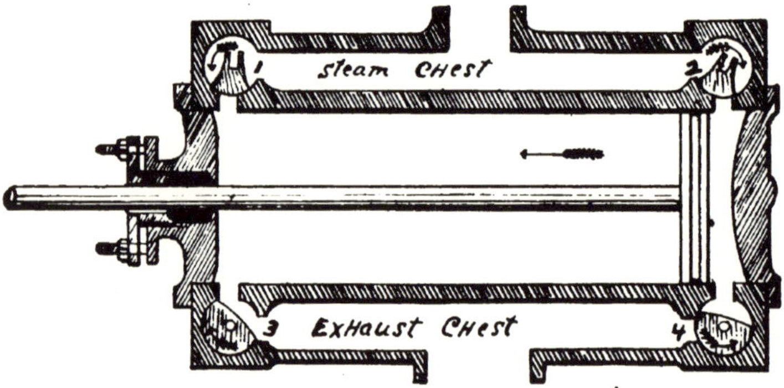

Fig. 121.

Fig. 121 is a sectional view of the cylinder, steam, and exhaust-chests, and the valve-chambers of a Corliss engine. 1 and 2 are the steam-valves and 3 and 4 the exhaust-valves. The valves work in cylindrical chambers accurately bored out, the face of the valve being turned off to fit steam tight. They are what is termed rotative valves, that is, they receive a semi-rotary motion from the wrist-plate, which in turn is actuated by the eccentric.

Ans.—The stern-wheel engine, consisting of a pair of engines, one cylinder on either side of the boat, and directly connected to the shaft of the stern-wheel. Like the beam engine, the stroke is long in proportion to the cylinder diameter.

Ques. 478.—Are these engines simple or compound?

Ans.—In former years simple engines were used altogether, but the later types are compound, either tandem or cross-compound.

Ques. 479.—What styles of valves and valve-gears are in use on these engines?

Ans.—Poppet-valves, actuated by long cam-driven levers, are the most generally used. Other styles of valves, such as rotative valves, common slide and piston-valves, are also quite frequently used.

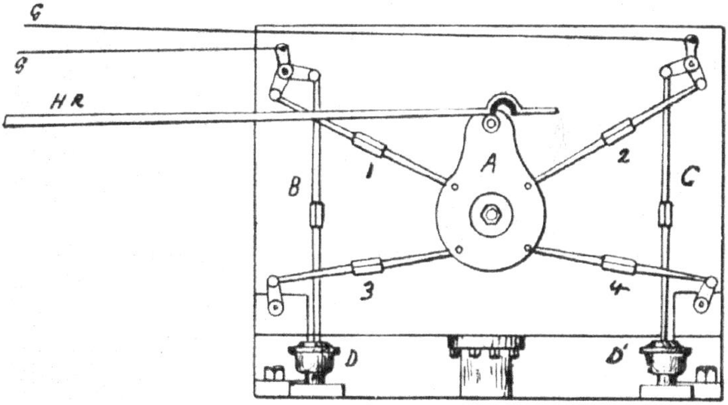

FIG. 122.

The valve-gear of a Corliss engine with a single eccentric is shown in Fig. 122. The connections of the exhaust-valves with the wrist-plate are positive, and the travel of these valves is fixed, being a constant quantity, but the connections of the steam-valves with the wrist-plate are detachable, being under the control of the governor.

Ques. 480.—What is meant in speaking of a four-valve engine?

Ans.—An engine having two steam-valves and two exhaust-valves located near each end of the cylinder.

Ques. 481.—What type of four-valve engine has met with great favor since its introduction?

Ans.—The Corliss engine, invented by Mr. Geo. H. Corliss, of Providence, R. I.

Ques. 482.—What advantage does the four-valve engine possess over the single-valve type?

Ans.—The advantage that each valve may be adjusted to a certain degree independently of the others, the steam-valves for admission and cut-off and the exhaust-valves for compression and release.

Ques. 483.—What is one of the oldest forms of valve, and one that is still used extensively, especially on marine engines?

Ans.—The D slide-valve.

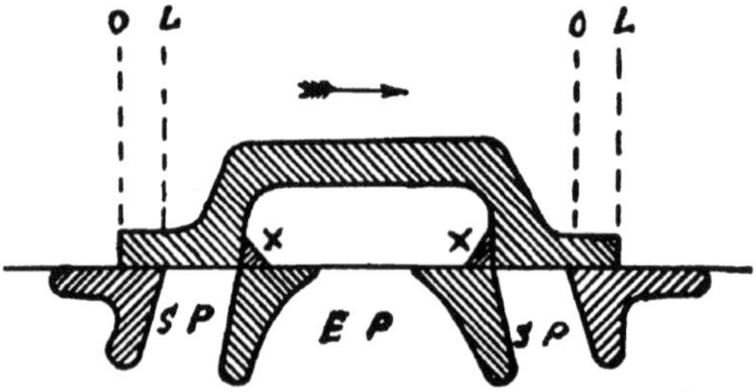

FIG. 123.

Fig. 123 represents a slide-valve at mid-travel. S P—S P are the steam-ports and E P is the exhaust-port: the projections marked x at each foot of the arch inside the valve represent inside lap, and may be added to or taken from the inside edges of the valve, according as more or less compression is desired. The dotted lines O L—O L represent outside lap.

Ques. 484.—What are the functions of the slide-valve?

Ans.—It controls the admission, expansion and release of the steam and the closure of the exhaust.

Ques. 485.—Upon what does the development of the full power of the engine and its efficient and economical use of steam, as well as its regular and quiet action, largely depend?

Ans.—Upon the correct adjustment of its valve or valves.

Ques. 486.—How is the slide-valve fitted to the cylinder?

Ans.—The slide-valve has a flat face and it works on the corresponding flat face of the cylinder. In the cylinder face there are three passages called ports, the two smallest, called steam-ports, leading to each end of the cylinder, and the larger one, called the exhaust-port, eading to either the receiver, condensor, or the atmosphere, as the case may be. The valve is contained in a steam-tight chest or casing, either cast with the cylinder,

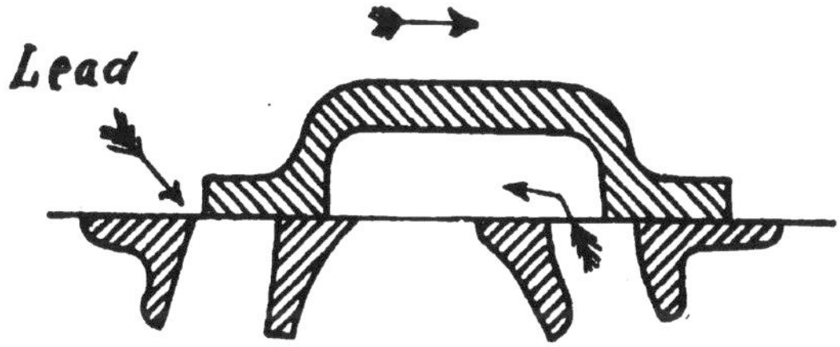

FIG. 124.

Fig. 124 shows the slide-valve in the position of lead—exhaust opening has also occurred at the opposite end of the cylinder. The arrows show the course of the steam, also the direction in which the valve is traveling.

or bolted to it. This casing or valve-chest is filled with live steam while the engine is working.

Ques. 487.—How must the slide-valve be constructed, in order that it may properly perform the four important functions of admission, cut-off, release, and exhaust closure?

Ans.—It must have lap and lead.

Ques. 488.—What is lap?

Ans.—Lap is the amount that the ends of the valve project over the edges of the ports when the valve is at mid-travel.

Ques. 489.—What is steam lap, or outside lap?

Ans.—The amount that the end of the valve projects over the outside edge of the steam-port.

Ques. 490.—What is inside or exhaust lap?

Ans.—The lap of the inside or exhaust edge of the valve over the inside edge of the port.

Ques. 491.—What is lead?

Ans.—The amount that the steam port is open when the piston is just commencing its stroke. This is the instant of admission.

Ques. 492.—When is the instant of cut-off?

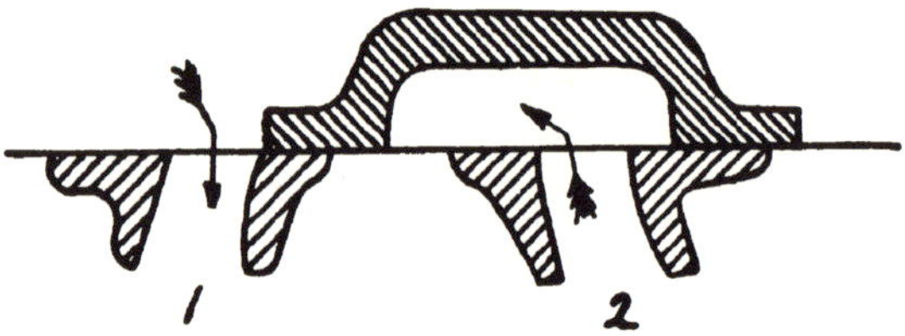

FIG. 125.

Fig. 125 shows the slide-valve at the end of its travel—full port opening.

Ans.—When the admission of steam to the cylinder is stopped by the steam edge of the valve closing the steam-port and the piston is pushed the balance of the stroke by the expansion of the steam admitted before cut-off occurred.

Ques. 493.—When is the instant of compression?

Ans.—When the two inside or exhaust edges of the valve coincide with the inner edges of the ports, the piston being near the end of its stroke and the valve at mid-travel.

Ques. 494.—When is the instant of release?

Ans.—When the inner edge of the valve commences to open the steam-port to the exhaust-passage.

Ques. 495.—What is the advantage gained by compression?

Ans.—A portion of steam is confined ahead of the piston, thus forming an elastic cushion to absorb the momentum of the piston and other moving parts connected with it and bring all to rest quietly at the end of the stroke.

Ques. 496.—How may this compression be increased or diminished?

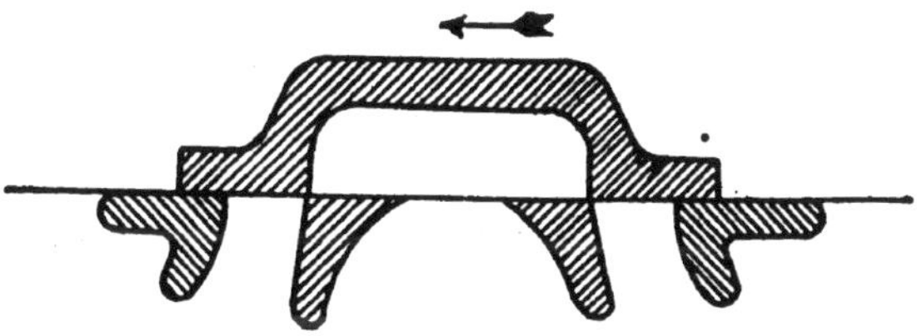

FIG. 126.

Fig. 126 illustrates the instant of cut-off. The valve is now traveling in the opposite direction.

Ans.—By adding to or taking away from the inside lap of the valve.

Ques. 497.—What is the object of giving a valve lead?

Ans.—The effect of lead is to cause the engine to be quick and not to lag at the beginning of the stroke. The live steam admitted through the lead opening also assists in forming a cushion for the piston at the end of the stroke.

Ques. 498.—Do the principles governing the adjustment and action of the slide-valve necessarily have to be applied in the adjustment and action of rotative, piston, and other forms of valves for controlling the distribution of steam in the cylinders of engines?

Ans.—They do. The same general principles apply in all cases.

Ques. 499.—How is motion generally imparted to the slide-valve or other types of valves?

Ans.—By means of an eccentric, which is simply a circular cast-iron or cast-steel sheave having a hole bored in it eccentrically with its own circumference, and large enough to permit of its being fitted on the engine shaft. The eccentric-sheave is either keyed on the shaft or held

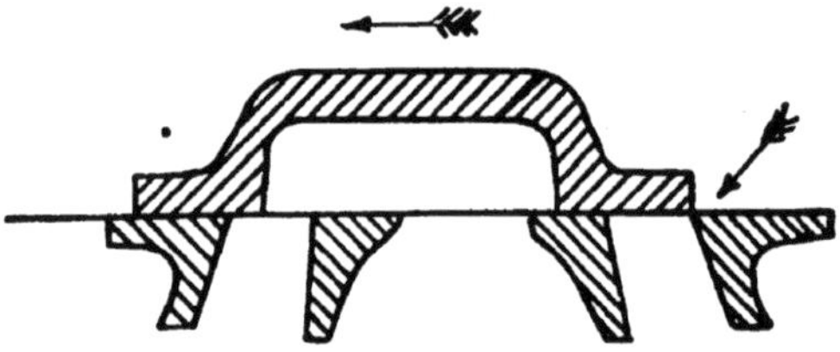

Fig. 127.

Fig. 127 shows the slide-valve at the instant of compression.

in its place by set-screws, and therefore revolves with the shaft. On the circumference of the eccentric, which is of sufficient width to present a good bearing surface, a ring, called the eccentric-strap, works, and attached to this ring is the eccentric-rod, which is either directly connected to the valve-rod, or valve-stem, or else imparts motion to the valve through the agency of a rocker-arm, and in many engines a link motion is used. The center of revolution of the eccentric being several inches apart from its center of formation, will, when the sheave revolves with the shaft, cause the eccentric to convert the rotary motion into a reciprocating motion, which through the agency of the rod is imparted to the valve or valves.

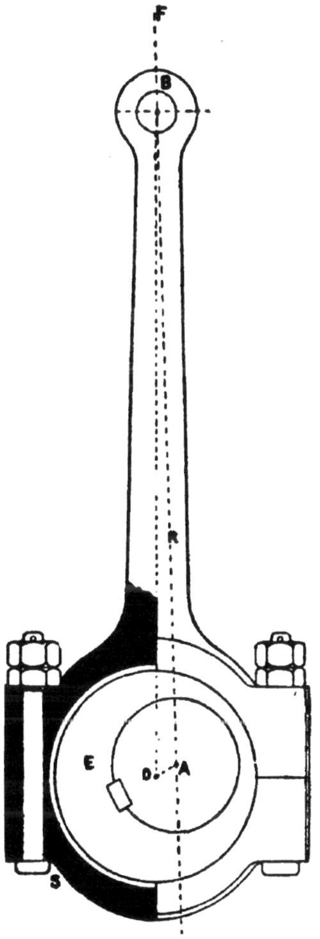

FIG. 128.

Fig. 128 shows an eccentric with its strap and rod. E is the sheave, the center of which is shown at D. A is the center of the shaft. The distance A D represents the throw of the eccentric and twice that distance equals the travel of the end B of the rod along the line A F. S is the eccentric strap.

Ques. 500.—What is meant by the throw of an eccentric?

Ans.—The distance between the center of the eccentric-sheave and the center of the crank-shaft. This distance is also called the radius of eccentricity.

Ques. 501.—What is meant by eccentric position?

Ans.—The location of the highest point of the eccentric relative to the center of the crank-pin, expressed in degrees.

Ques. 502.—What is angular advance?

Ans.—The distance that the high point of the eccentric is set ahead of a line at right angles with the crank, in other words, the lap angle plus the lead angle.

Ques. 503.—If a valve had neither lap nor lead, what would be the position of the high point of the eccentric relative to the crank?

Ans.—It would be on a line exactly at right angles with the crank, as, for instance, the crank being at 0 degrees, the eccentric would stand at 90 degrees.

Ques. 504.—How is the reversing of modern marine engines usually effected?

Ans.—By means of the link motion, using two eccentrics.

Ques. 505.—How many varieties of links are in use?

Ans.—Three, the slotted link, the solid-bar link and the double-bar link.

Ques. 506.—Describe the slotted link.

Ans.—It is a curved bar with a slot cut in it, in which the link-block is fitted. This link-block is attached to the valve-rod by a pin, about which an oscillating motion of the block occurs. Two projectoins are formed on one side of the link, to which the eccentric rods are connected.

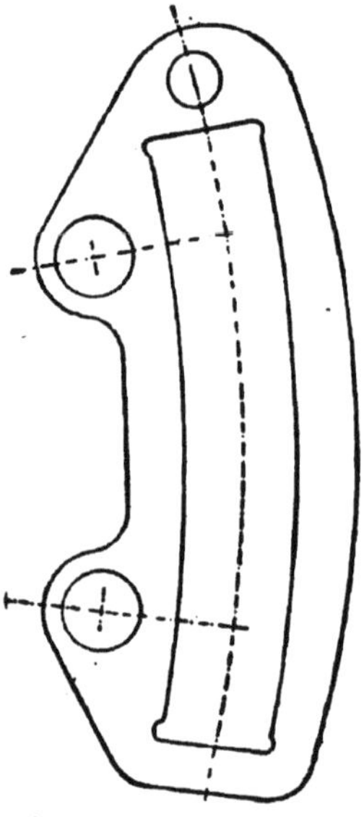

Fig. 129. Slotted Link.

Ques. 507.—Describe the solid-bar link.

Ans—The solid-bar link consists of a simple, curved, rectangular bar, with eyes formed at each end, for connecting the eccentric-rods. The solid bar passes through the block.

Ques. 508.—What is the general plan of the double-bar link?

Ans.—It consists of a pair of curved steel bars joined

at the ends and kept a certain distance apart by distance pieces. Projecting pins are formed on the link-bars, two on each side, for the attachment of the eccentric-rods. The ends of the eccentric-rods are forked and contain each two adjustable bearings, which embrace the pins on each side of the link. The link-block is a steel or iron pin, sliding between the bars, and having projections on each

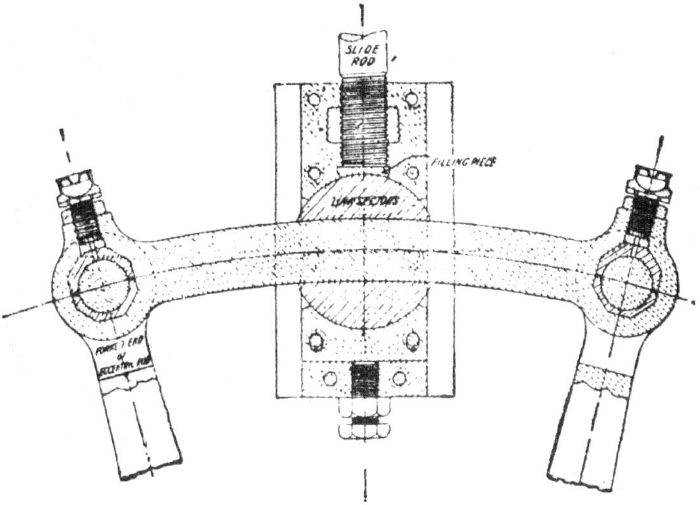

FIG. 130. SOLID BAR LINK.

side which embrace the link-bars and through which these bars slide, on adjustable gun-metal liners.

Ques. 509.—Why is it necessary that the link should be curved?

Ans.—For the reason that it is used not only for reversing the engine, but also for working steam expansively, and therefore its shape must be such that when the block is in any intermediate position the center of the travel of the valve will always be constant, otherwise the distribution of the steam to the two ends of the cylinder would not be evenly divided.

Ques. 510.—What is the slip of the link?

Ans.—A slight oscillating movement of the link on its block.

Ques. 511.—What is it that fixes the curvature of the link?

Ans.—The length of the eccentric-rod; that is, the curve of the link is a circular arc, of a radius equal to the

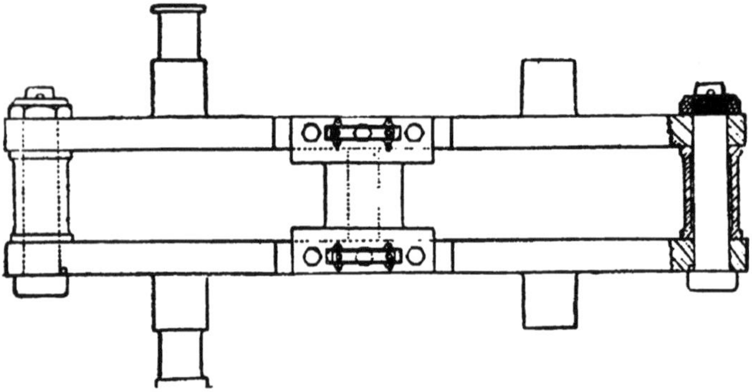

FIG. 131. PLAN VIEW OF DOUBLE-BARRED LINK.

distance between the center of the eccentric and center of the pin at the end of the rod.

Ques. 512.—Is there a type of reversing valve-gear that employs but one eccentric?

Ans.—There is, viz., the Marshall radial valve-gear.

Ques. 513.—It there a type of reversing valve-gear in which eccentrics are dispensed with?

Ans.—There is, viz., the Joy valve-gear, by which the motion of the valve is derived from the connecting rod, through the medium of a vibrating link, one end of which is jointed to the connecting rod while the other end is constrained to move in a horizontal or vertical direction by the action of a radius rod. This motion is horizontal if the engine is a vertical engine, or vertical if the engine is horizontal. One end of another rod works on a pin in the vibrating link and near the other end of this rod is a fulcrum carried by a pin attached to sliding blocks on each side, working in sectors, which are carried by the reversing shaft. Motion is communicated to the valve

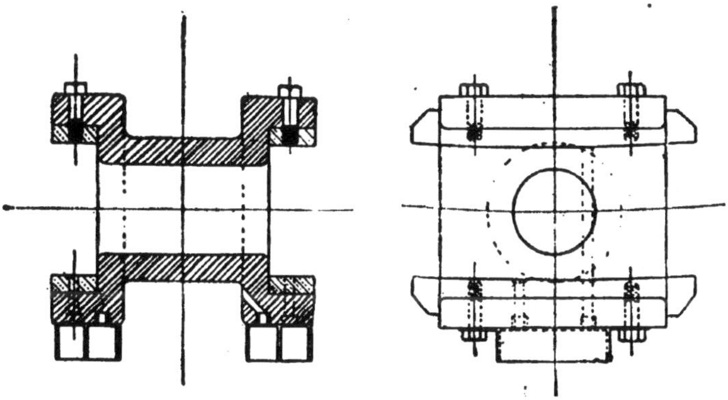

FIG. 132.

Fig. 132 shows details of the construction of the link-block for a double-barred link.

from a point in the last-mentioned rod beyond the fulcrum carried by the sectors attached to the reversing shaft.

Ques. 514.—How is the forward or backward movement of the engine effected with the Joy valve-gear?

Ans.—By inclining the sector on one or the other side of the center line of the reversing shaft and the point of cut-off and consequently the amount of expansion depends upon the amount of the inclination, the central position being mid-gear. The reversing arm moves these sectors to the required position. In large marine engines the reversing mechanism is operated by a small starting engine. On locomotives and small engines it is operated by hand.

Ques. 515.—Mention one of the advantages possessed by the Joy valve-gear, over the double eccentric and link motion.

Ans.—By this gear a constant lead is secured for all linked-up positions.

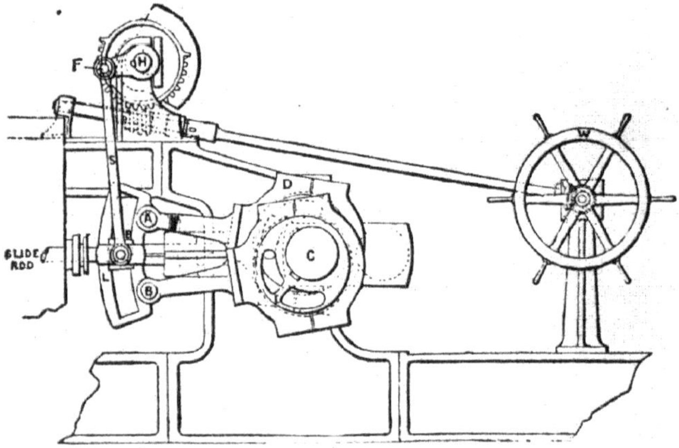

FIG. 133. STEVENSON LINK MOTION FOR MARINE ENGINE.

On the crank-shaft C there are keyed two eccentrics, one in the position to give ahead motion and the other in the position for astern motion. The eccentric rods are of equal length, and their ends are attached by working joints to the opposite ends of a curved link, L.

Ques. 516.—Is the Joy valve-gear much in use?

Ans.—It is applied to a large number of marine engines and locomotives.

Ques. 517.—How is the lifting valve-gear of the marine beam engine actuated?

Ans.—By curved cams keyed to a transverse shaft Four cams are fitted, two for steam and two for exhaust

Ques. 518.—How is the oscillating movement imparted to the transverse shaft?

Ans.—From rocker-arms, one on each end of the

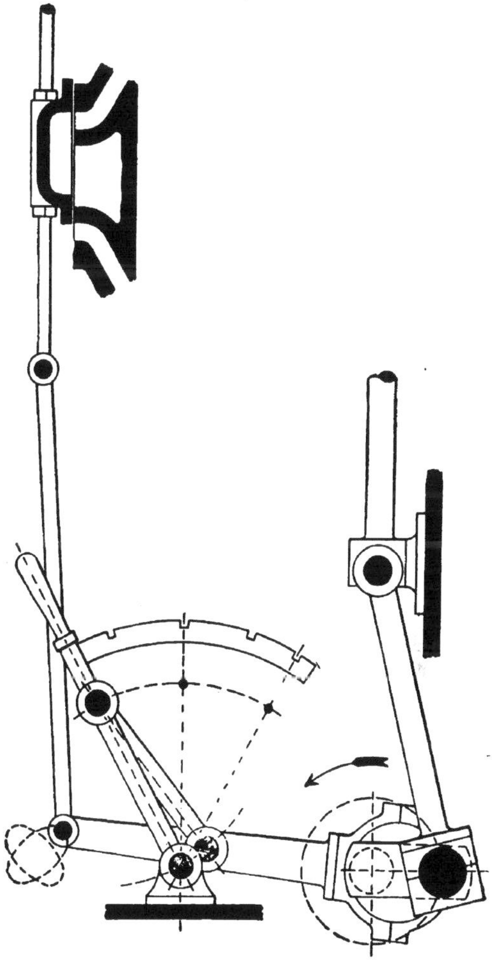

FIG. 134.

The Marshall Valve-gear. The single eccentric, turning with the crank, works the valve through a pivoted arm. The movement of the engine may be stopped or reversed by sliding the hand-lever on the notched quadrant. The loop paths show the movement of the valve rod-pin and also of the valve in vertical directions for ahead or backing motion.

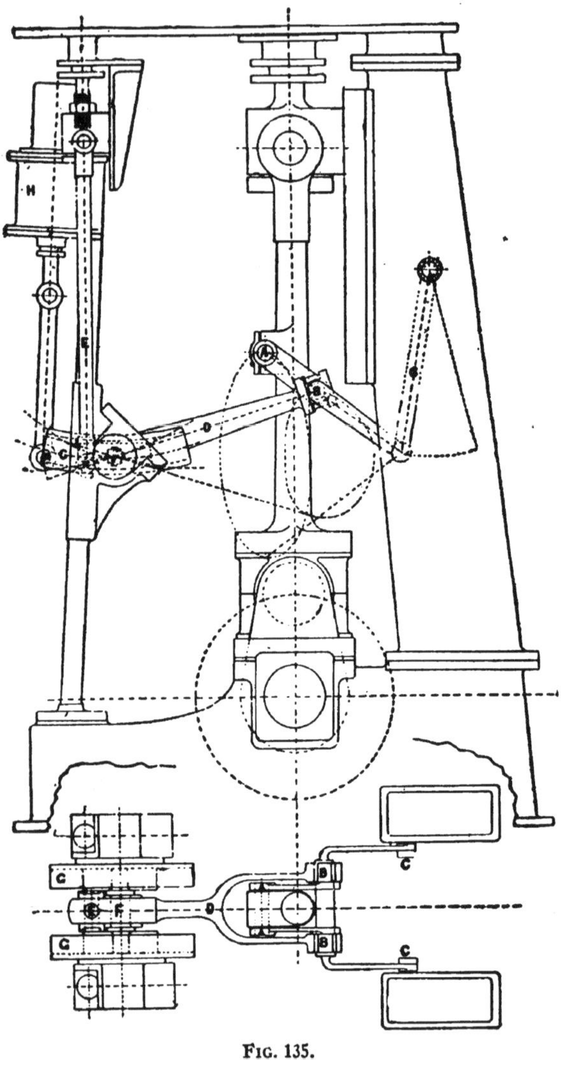

FIG. 135.

Fig. 135 shows an elevation and plan of the latest arrangement of Joy's valve-gear applied to a vertical engine. In this gear eccentrics are dispensed with, and the movements of the slide-valve obtained from the connecting rod. The vibrating link B, jointed to the connecting rod at A, has one end constrained to move horizontally by the action of the radius rod C. One end of another rod, D, works on a pin in the vibrating link B; near the other end is a fulcrum carried by a pin F, attached to sliding blocks on each side working in sectors G, which are carried by the reversing shaft, the center line of the sector passing through the center of the reversing shaft. From D the motion is communicated to the slide-valve rod by means of the link E, attached to a point K in the rod D beyond the fulcrum F.

The forward or backward movement of the engine is governed by inclining the sector on one or the other side of the horizontal center line, and the amount of expansion depends on the amount of the inclination, the exactly central or horizontal position being 'mid-gear.' The reversing arm F R moves these sectors to the required position, and its extremity R is connected to the starting engine H. The paths of the point A in the connecting rod, and also of the point B in the vibrating link, as the engine revolves, are indicated by dotted lines, as are also the extreme positions of the sector center lines for ahead and astern working respectively. The gear as drawn is in the stop position. By this gear a constant lead is secured for all linked-up positions, since when the piston is at the top or bottom of the stroke the pin F co-incides with the center of the reversing shaft, so that in this position any movement of the sectors does not affect the position of the slide-valve. The up and down motion of the point B therefore gives a constant movement of the valve equal to the *lap plus the lead*, while the horizontal motion sliding the block to and fro in the sectors adds the amount required for steam opening, this amount increasing with the angle of the sector to the horizontal.

transverse shaft, the pins of which are engaged by hooks in the eccentric-rods.

Ques. 519.—How are the poppet-valves of the Western river boat actuated?

Ans.—By cams very similar to those on the beam engine.

Ques. 520.—Describe the construction and action of a double-ported slide-valve.

Ans.—A double-ported slide-valve acts in a manner similar to a single-ported valve in the admission of steam, but in addition there is what is practically an inner valve, to which steam is admitted through passages formed in the body of the valve. There are also two inner ports in addition to the two ports at the ends of the cylinder, and these inner ports or passages also lead to and are in connection with the end ports.

Ques. 521.—What is the object in using a double-ported valve?

Ans.—To reduce the travel of the valve, which in large engines would be too great with a single-ported valve.

Ques. 522.—Are treble-ported valves used to any large extent?

Ans.—They are, on large marine engines, their action being on the same general principles as the double-ported slide-valve.

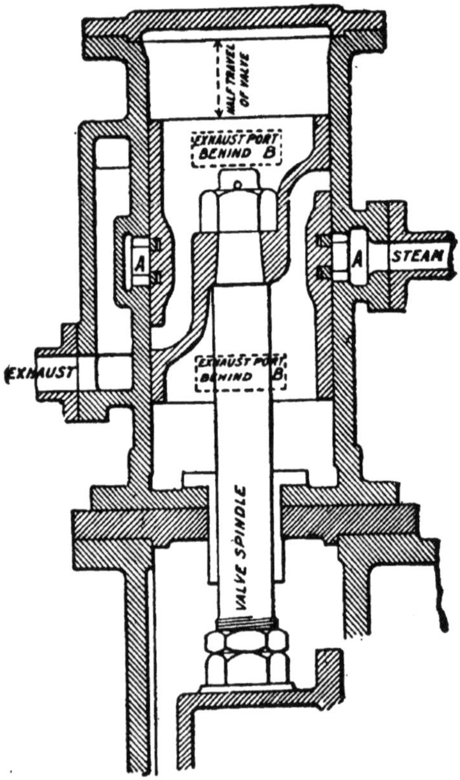

FIG. 136. JOY'S ASSISTANT CYLINDER.

This consists of a small cylinder and steam-piston attached to the valve-spindle. The cylinder has a central inlet for steam, A, and two exhaust-ports, B, one for each end, leading to a common exhaust-pipe, and the piston is so constructed that by its motion the operations of steam admission, cut-off, release and compression are performed on each side of the piston. The apparatus is, therefore, a small engine which exercises a force on the valve to move it up or down, and cushions steam at each end to absorb the momentum forces. These assistant cylinders give diagrams similar to that of an ordinary engine; they exert from 15 to 25 I. H. P. each for the sizes fitted in marine engines, and the amount of power developed can be adjusted by means of a valve on the steam-pipe. If the main valve be linked in, the assistant cylinder is also automatically similarly affected.

Ques. 523.—How is the pressure of the steam on the back of a large, flat slide-valve lessened and relieved?

Ans.—By relief packing rings.

Ques. 524.—How are relief packing rings fitted?

Ans.—They are sometimes fitted on the back of the valve, but are generally fitted on the valve-chest cover, and are pressed out by springs so as to work steam-tight on a planed surface, either on the back of the valve or, on the inside of the cover, thus reducing the area on which the steam-pressure can act. The space inside the packing ring is connected to the condensor or the receiver of the succeeding engine.

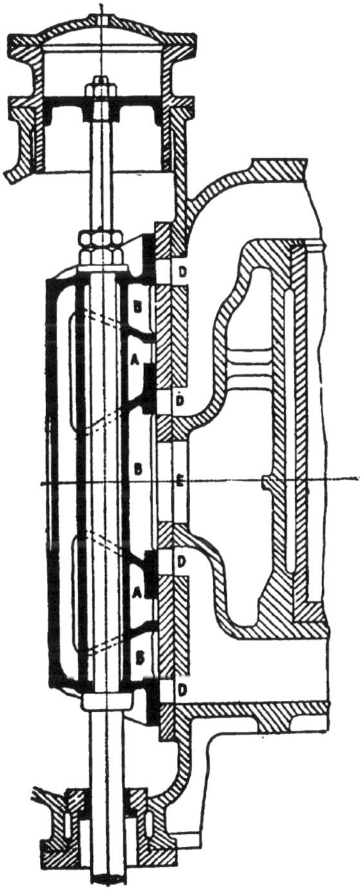

FIG. 137. DOUBLE-PORTED VALVE.

Ques. 525.—Do relief rings work in a satisfactory manner?

Ans.—They do not, as a general thing, being troublesome to make efficient, and they also are difficult of adjustment.

Ques. 526.—What form of slide-valve has been found to give good satisfaction, especially on the high and intermediate cylinders of large marine engines?

Ans.—The piston slide-valve, consisting of two pistons connected together and working steam-tight in cylindrical chambers that contain the steam-ports. The face of each of the pistons corresponds to the face of the single-ported slide-valve, and performs the same functions.

Ques 527.—What means are provided in large vertical engines for preventing the weight of the slide-valve, rod, and link gear from bearing upon the eccentric?

Ans.—Balancing pistons, working in small steam-cylinders in the top end of the valve-casing.

Ques 528.—What is meant by setting the valve, or valves of an engine?

Ans.—The adjustment and securing of the slide-valve in its proper position on the rod so as to secure the correct distribution of the steam in the cylinder. This also includes the fixing of the eccentric in its correct position on the shaft.

Ques. 529.—What is the first, or one of the first, moves in valve-setting?

Ans.—The rods and gear are first coupled together, and the crank is placed on the dead center.

Ques. 530.—What is the meaning of the expression dead center?

Ans.—An engine is on the dead center when the crank is in line with the piston-rod, that is, when the centers of the crank-shaft, crank-pin, and cross-head pin are exactly in line, so that the pressure of the steam on the piston exerts no turning moment on the shaft, but produces only direct thrust, subjecting the shaft to bending action only.

Ques. 531.—With the engine on the dead center and

the rods and valve-gear all coupled up, what is the next move in valve-setting?

Ans.—The slide-valve, by means of screws and nuts on the valve-rod, is fixed in the proper position to give the required lead for the corresponding end of the cylinder. The shaft is then turned around until the crank is on the opposite dead point and the lead of the valve for that end of the cylinder is measured. If the amounts of lead at the opposite ends are different, the position of the slide-valve on the rod should be adjusted by means of the nuts and screws, until the leads are either equal, or differ by the desired amount. The valve should then be permanently secured on the rod, so that its position may not alter. This is called equalizing the lead.

Ques. 532.—If, after having gotten the lead equalized, it is found that there is too much or too little, how may it be decreased, or increased without altering the position of the valve on the rod?

Ans.—To decrease the lead, reduce the angular advance of the eccentric, and to increase the lead it is necessary to increase the angular advance.

Ques. 533.—What is the rule generally observed regarding the lead on large vertical engines?

Ans.—In vertical engines, owing to the weight of the moving parts, the lead on the lower end is generally made slightly greater than the lead on the upper end, and more exhaust lap is allowed. In such cases the valve is set on the rod to give the required difference between the two leads. Then, if the lead be too great or too small at both ends, the required change may be made by moving the eccentric ahead or back on the shaft.

Ques. **534.**—Is the position of the eccentric on the shaft necessarily fixed on all types of engines?

Ans.—It is not. Many high-class stationary engines

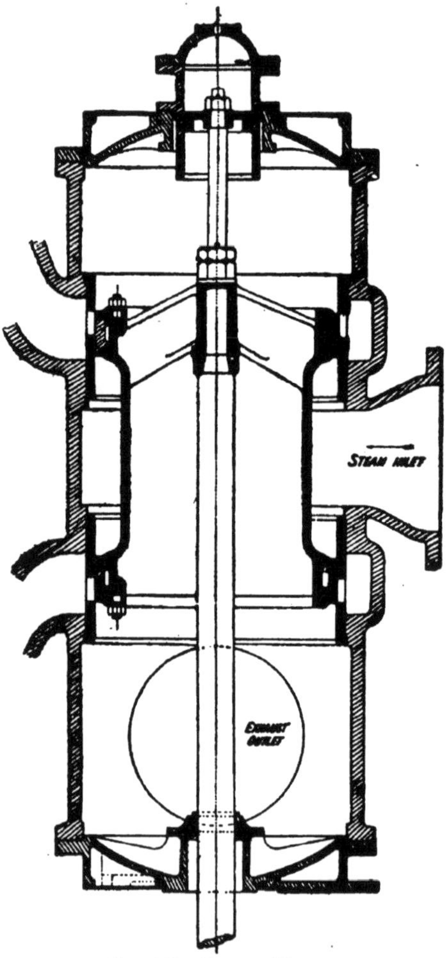

FIG. 138. PISTON VALVE.

are fitted with isochronol or inertia governors, which control the position of the eccentric and vary the point of cut-off according as the load on the engine is light or heavy, thus maintaining a regular speed.

Ques. 535.—What types of valves are used with isochronol governors?

Ans.—Slide-valves of various patterns; box-valves, in which the steam passes through the valve; piston valves, in which the steam either passes through or around the ends of the valve.

Ques. 536.—In all types of reciprocating engines the same factors affecting the distribution of the steam are present. What are they?

Ans.—Outside lap, affecting admission and cut-off, and inside lap, affecting release and compression.

Ques. 537.—How are these factors distributed in the four-valve type of engine?

Ans.—They are distributed among the four valves, each valve performing its own particular function in the distribution of the steam for the end of the cylinder to which it is attached.

Ques. 538.—What advantage is there connected with setting the valves of a four-valve engine, as compared with a single valve?

Ans.—Each valve may be adjusted to a certain degree independently of the others, thus, for instance, the steam-valves of a Corliss engine may be adjusted to cut off the steam at any point from the beginning up to one-half the stroke, without in the least affecting the release or compression, because these latter events are controlled by the exhaust-valves.

Ques. 539.—What is the first requisite in setting the valves of a Corliss engine?

Ans.—To place the crank on the dead center.

Ques. 540.—What is the next move?

Ans.—To adjust the length of the hook-rod, if it is adjustable; if not, then the length of the eccentric-rod, so that the wrist-plate will vibrate equal distances each

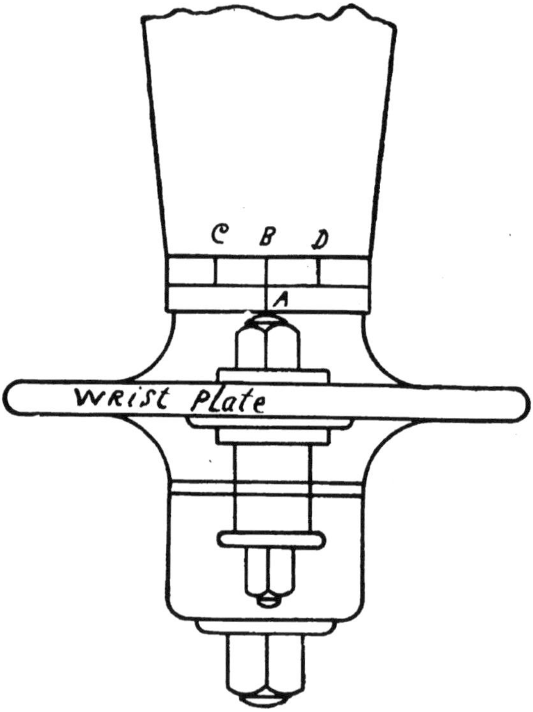

FIG. 139. WRIST PLATE OF CORLISS ENGINE.

way from its central position, which is marked on top of the hub.

Ques. 541.—How should the rocker-arm, that carries the eccentric-rod, and hook-rod be adjusted?

Ans.—The length of the eccentric-rod should be such

that the rocker-arm will vibrate equal distances each way from a vertical position.

Ques. 542.—How may the vibration of the wrist-plate and rocker-arm be tested?

Ans.—By connecting the eccentric-rod and the hook-rod in their proper places, and turning the loose eccentric around on the shaft in the direction the engine is to run.

Ques. 543.—Having gotten these important adjustments correctly made, what is the next step in setting Corliss valves?

Ans.—Remove the back bonnets from the four valve

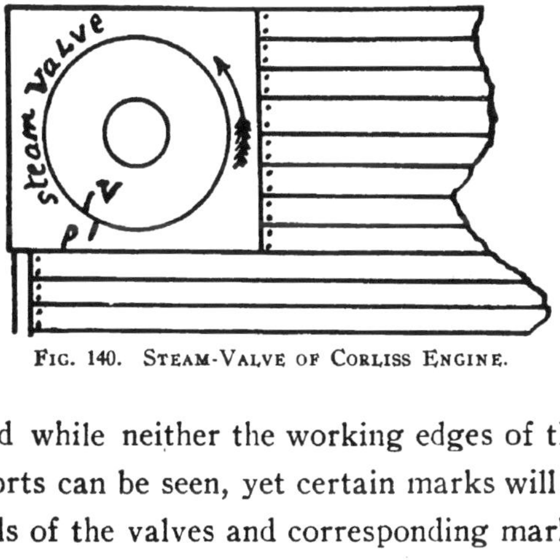

FIG. 140. STEAM-VALVE OF CORLISS ENGINE.

chests, and while neither the working edges of the valves nor the ports can be seen, yet certain marks will be found on the ends of the valves and corresponding marks on the faces of the chests, which serve as a guide in setting the valves.

Ques. 544.—Having removed the bonnets and found the marks, what is to be done next?

Ans.—Temporarily secure the wrist-plate in its central position by tightening one of the set-screws on the eccentric. Then connect the valve-rods to the wrist-plate and to the small crank-arms attached to the ends of the valves, adjusting their lengths so that the steam-valves will have from ¼ to $\frac{9}{16}$ inch lap, and the exhaust valves from $\frac{1}{32}$ to $\frac{3}{16}$ inch opening.

Ques. 545.—In adjusting the steam-valves, what particular detail should be carefully noted?

Ans.—The direction in which the valves turn to open should be noted. In most Corliss engines the arm of the small crank to which the valve-rod is connected, extends

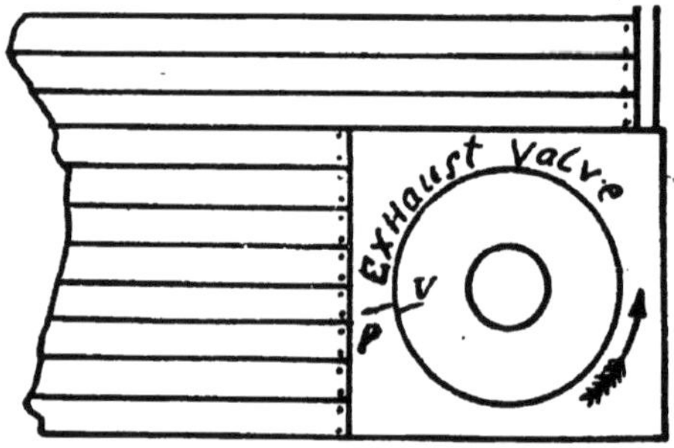

FIG. 141. EXHAUST-VALVE OF CORLISS ENGINE.

downwards from the valve-stem. This will cause the valve to move towards the wrist-plate in opening.

Ques. 546.—After the valve-rods have been properly adjusted as to length, what is the next move?

Ans.—Place the engine on the dead center—either center will do—and move the eccentric around on the shaft in the direction the engine is to run, until the eccentric is far enough ahead of the crank to allow the steam-valve for that end of the cylinder the proper amount of lead opening, which will vary according to the size of the engine. Then tighten the eccentric set screws and turn the engine around to the opposite center and note whether the lead is the same on both ends.

Ques. 547.—In case there is a difference in the lead for the two ends, how may it generally be equalized?

Ans.—By slightly altering the length of one of the valve-rods.

Ques. 548.—What is the next point to receive attention, in setting Corliss valves?

TABLE 8

LAP AND LEAD OF CORLISS VALVES

Size of Engine.	Lap of Steam Valve.	Lead Opening of Steam Valve.	Lead Opening of Exhaust Valve.
12 inches	$\frac{1}{4}$ inch	$\frac{1}{32}$ inch	$\frac{1}{32}$ inch
14 "	$\frac{5}{16}$ "	$\frac{1}{32}$ "	$\frac{1}{32}$ "
16 "	$\frac{5}{16}$ "	$\frac{1}{16}$ "	$\frac{1}{32}$ "
18 "	$\frac{3}{8}$ "	$\frac{1}{16}$ "	$\frac{1}{16}$ "
20 "	$\frac{3}{8}$ "	$\frac{1}{16}$ "	$\frac{1}{16}$ "
22 "	$\frac{3}{8}$ "	$\frac{1}{16}$ "	$\frac{1}{16}$ "
24 "	$\frac{7}{16}$ "	$\frac{3}{32}$ "	$\frac{3}{32}$ "
26 "	$\frac{7}{16}$ "	$\frac{3}{32}$ "	$\frac{3}{32}$ "
28 "	$\frac{7}{16}$ "	$\frac{3}{32}$ "	$\frac{3}{32}$ "
30 "	$\frac{1}{2}$ "	$\frac{3}{32}$ "	$\frac{1}{8}$ "
32 "	$\frac{1}{2}$ "	$\frac{3}{32}$ "	$\frac{1}{8}$ "
34 "	$\frac{1}{2}$ "	$\frac{1}{8}$ "	$\frac{1}{8}$ "
36 "	$\frac{1}{2}$ "	$\frac{1}{8}$ "	$\frac{1}{8}$ "
38 "	$\frac{9}{16}$ "	$\frac{1}{8}$ "	$\frac{3}{16}$ "
40 "	$\frac{9}{16}$ "	$\frac{1}{8}$ "	$\frac{3}{16}$ "
42 "	$\frac{9}{16}$ "	$\frac{1}{8}$ "	$\frac{3}{16}$ "

Ans.—The adjustment of the lengths of the rods extending from the governor to the releasing mechanism, so that the valves will cut off at equal points in the stroke.

Ques. 549.—How is this adjustment accomplished?

Ans.—By raising the hook-rod clear of the wrist-plate pin and with the bar provided for the purpose, move the

14

wrist-plate to either one of its extreme positions, as shown by the marks on the hub, and, holding it in this position, adjust the length of the governor-rod for that steam-valve (which will then be wide open) so that the boss or roller which trips the releasing mechanism is just in contact, or within $\frac{1}{32}$ inch of it. Then move the wrist-plate to the other extreme of its travel and adjust the length of the other rod in the same manner.

Ques. 550.—How may the accuracy of this adjustment be tested?

Ans.—Raise the governor-balls to their medium position, or about where they would be when the engine is running at its normal speed, and block them there. Then having again connected the hook-rod to the wrist-plate, turn the engine around in the direction in which it is to run, and when the valve is released by the trip, measure the distance upon the guide that the cross-head has traveled from the end of the stroke. Now continue to turn the engine in the same direction until the other valve is released, and measure the distance that the cross-head has traveled from the opposite end of the stroke, and if these two distances are the same, the cut-off is equalized. If there is a difference, lengthen one rod and shorten the other until the point of cut-off is the same for both ends. The lengths of the dash-pot rods should also be adjusted, so that when the plunger is at the bottom of the dash-pot the valve-lever will engage the hook. The lock-nuts on all rods should then be securely tightened.

CHAPTER VII

CONDENSERS—AIR-PUMPS—SEA-WATER

Ques. 551.—What is the average composition of sea-water?

Ans.—Sea-water contains about $\frac{1}{32}$ part of its weight of solid matter, of which common salt (sodic chloride) is the principle constituent. The average composition of the solid matter in sea-water may be taken as follows:

Sodic chloride, or common salt	76	per cent
Magnesic chloride	10	"
Calcic sulphate, or gypsum	5	"
Magnesic sulphate	6	"
Carbonate of lime, and organic matter	3	"

Ques. 552.—Does the common salt in sea-water cause much trouble for the marine engineer?

Ans.—It does not, for the reason that it remains soluble in water at all temperatures, and there is no deposit of salt, except under extreme circumstances.

Ques. 553.—What is the principal scale-forming ingredient in sea-water?

Ans.—Sulphate of lime, or calcic sulphate. Deposit is also formed by sulphate of magnesia, although it is less objectionable than the lime deposit.

Ques. 554.—At what temperature does the sulphate of lime become insoluble in water and form a deposit on the boiler plates?

Ans.—At a temperature of 280 degrees to 295 degrees

Fahrenheit, corresponding to a pressure of 35 to 45 pounds pressure of steam by the gauge. As the temperature of the water rises, the other sulphates become insoluble, and at 350 degrees Fahrenheit, or 120 pounds gauge-pressure, sea-water is incapable of holding any sulphates in solution.

Ques. 555.—What other cause, besides a high temperature, tends to precipitate these salts?

Ans.—Increase of density, caused by evaporation of the water, even if the temperature remains about 212 degrees Fahrenheit. Sulphate of calcium is thus deposited at a density of $\frac{3}{32}$. Common salt does not crystallize out until a density of about $\frac{8}{32}$ is reached.

Ques. 556.—When was it possible to use sea-water for feeding boilers?

Ans.—In the early days of marine engineering, when a low-pressure (35 to 45 pounds) was carried, and the jet condenser was used, in which the steam was exhausted into the condensing chamber, where it came into actual contact with and was condensed by a jet of cold sea-water. The feed-water for the boilers was drawn from this mixture of sea-water and condensed steam, consequently a large quantity of sea-water was sent into the boilers, but as the temperature was low and the density was not allowed to exceed $\frac{2}{32}$, the salts were held in solution fairly well.

Ques. 557.—How was the increase of density prevented?

Ans.—By blowing off a portion of the denser boiler-water at stated times, and making up the loss by admitting a larger quantity of salt water. This was termed "brining the boiler."

Ques. 558.—What led to the introduction of the surface condenser?

Ans.—With the advent of high pressures, it was found impossible to prevent the deposit of scale, and all of its attendant evils. It was therefore found necessary to condense the exhaust steam without bringing it into actual contact with the condensing water, hence the surface condenser was designed.

Ques. 559.—Mention two of the principal advantages gained by the use of the surface condenser.

Ans.—First, by its use fresh feed-water is obtained for the boilers; second, the condition of the condensing water is of no importance, as regards the feed-water so that, no matter whether it is salt, muddy, acid, or otherwise impure, pure water is always obtained for the boilers, provided the condenser is maintained in good condition and no leakage is allowed to occur.

Ques. 560.—What is the meaning of the word vacuum?

Ans.—That condition existing within a closed vessel during the absence of all pressure, including atmospheric pressure.

Ques. 561.—How is a vacuum measured?

Ans.—It is measured in inches of a column of mercury contained within a glass tube a little more than 30 inches in height, having its lower end open and immersed in a small open vessel filled with mercury. The upper end of the glass tube is connected with the vessel in which the vacuum is to be produced. When no vacuum exists, the mercury will leave the tube and fill the lower vessel. When a vacuum is maintained within the condenser, or other vessel, the mercury will rise in the glass tube to a height corresponding to the degree of vacuum. If the mercury rises to a height of 30 inches it indicates a perfect vacuum, which means the absence of all pressure within the vessel, but this condition is never realized in practice, the nearest approach to it being about 28 inches.

Ques. 562.—Is the mercurial vacuum-gauge used in every-day practice?

Ans.—For purposes of convenience it is not generally used, it having been replaced by the Bourdon Spring-gauge, although the mercury-gauge is used for testing.

Ques. 563.—What is the advantage, from a purely economic standpoint, in allowing the exhaust steam to pass into a condenser in which a vacuum is maintained rather than to allow it to exhaust into the open air?

Ans.—In a non-condensing engine, that is, an engine in which the exhaust steam passes into the open air, the pressure of the atmosphere, amounting to 14.7 pounds per square inch at sea-level, is constantly in resistance to the motion of the piston. Therefore the exhaust or terminal pressure can not fall below the atmospheric pressure and is generally from 2 to 5 pounds above it, caused by the resistance of bends, and turns in the exhaust pipe, or other causes which tend to retard the free passage of the steam. On the other hand, if the steam were allowed to exhaust into a condenser in which a vacuum of 25 inches is being maintained, the terminal pressure or back pressure in resistance to the forward motion of the piston would be but 2.5 pounds, and if a vacuum of 28 inches existed in the condenser there would be practically no back pressure, thus making available for useful work the 14.7 pounds of steam which in the non-condensing engine was required to overcome the resistance of the atmospheric pressure.

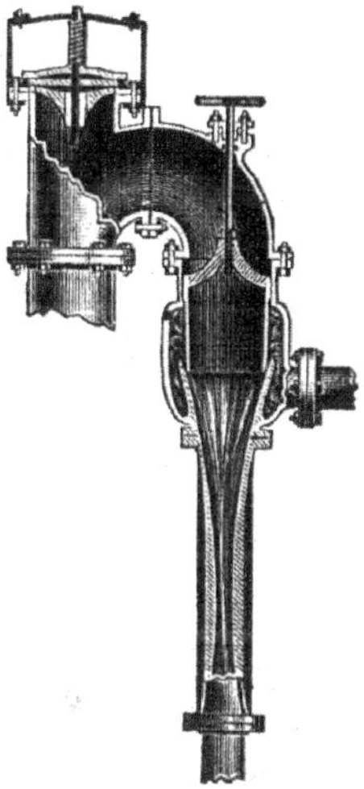

FIG. 142.
SIPHON CONDENSER.

Ques. 564.—Is it proper, then, to consider the vacuum in a condenser as power?

Ans.—The vacuum can not be considered as power at all. It occupies the anomalous position of increasing, by its presence, the capacity of the engine for doing work.

Ques. 565.—How is the vacuum in a condenser usually maintained?

Ans.—By a pump called an air-pump, although a partial vacuum can be produced by the mere condensation of the exhaust steam as it enters the condenser, by allowing a spray of cold water to strike it. The steam when it first enters the condenser drives out the air and the vessel is filled with steam at a low pressure, which, when condensed, occupies about 1,600 times less space than it did before being condensed, hence a partial vacuum is produced. The action of the siphon injector is based upon this principle.

Ques. 566.—Describe the construction and action of the siphon condenser.

Ans.—The siphon condenser is a form of jet condenser in which no air-pump is used. In this type of condenser the supply of condensing water is drawn from outside pressure, either from an overhead tank, or other source,

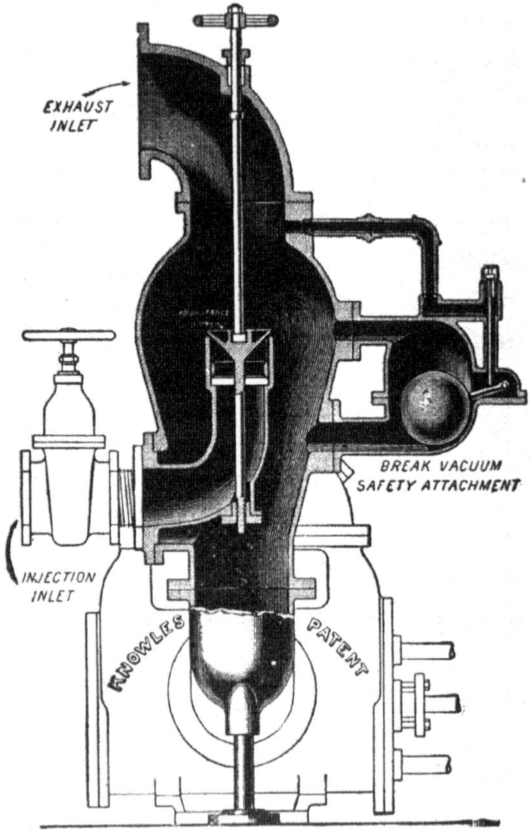

FIG. 143. KNOWLES JET CONDENSER.

and passing into an annular enlargement of the exhaust-pipe, is discharged downwards in the form of a cylindrical sheet of water, into a nozzle which gradually contracts. The exhaust steam, entering at the same time, is condensed and the contracting neck of the cone-shaped nozzle gradually brings the water to a solid jet and it rushes through the nozzle with a velocity sufficient to create a vacuum. This type of condenser can only be used where the discharge pipe has a perfectly free outlet.

Ques. 567.—Describe in general terms the construction and action of the jet condenser.

Ans.—The jet condenser is usually a vertical, cylindri-

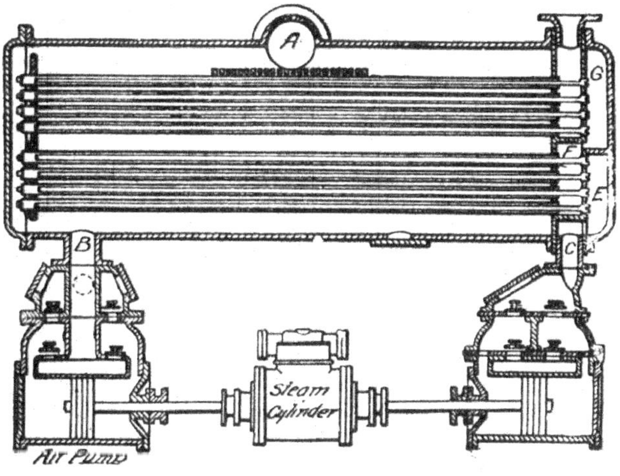

FIG. 144. SECTIONAL VIEW OF A SURFACE CONDENSER AND INDEPENDENT AIR AND CIRCULATING PUMPS.

cal, cast-iron vessel, made air-tight, and which receives the exhaust steam from the low-pressure cylinder. In modern plants, condenser-shells are often made of sheet steel in cylindrical shape, reenforced with stiffening rings. The exhaust steam enters at the top and the condensing water enters usually at the side, flowing in through the spraying nozzle, and, discharging through a large number of small holes, comes in contact with the steam in the form of spray, thus producing a quick condensation while falling to the bottom of the condenser, to be drawn off by the air-pump. A cock or valve is fitted in the injection pipe, for the purpose of regulating the supply of cooling water.

Ques. 568.—Why is an air-pump a necessary part of a reliable jet-condensing apparatus?

Ans.—The mixture of condensing water and condensed steam must be pumped away constantly, also the condensing water always contains a certain volume of air in solution, which may be liberated, either by boiling it or by reducing the pressure to which it is subjected. This air is liberated in the condenser, and if it is not pumped away regularly, it is liable to accumulate and spoil the vacuum.

Ques. 569.—How may the dimensions of a single-acting air-pump for a given sized engine be determined?

Ans.—In the solution of this problem, two factors must be considered: First, the total volume of the low-pressure cylinder; second, the density of the exhaust steam. The volume of the air-pump cylinder is then found by the following rule: Multiply the volume of the low-pressure cylinder in cubic feet by 3.5, and divide the product by the number of cubic feet contained in 1 pound weight of exhaust steam at the pressure at which it enters the condenser. This rule applies only to jet condensers.

Ques. 570.—Describe the construction and action of the surface condenser.

Ans.—The surface condenser, like the jet condenser,

is an air-tight iron or steel vessel, either cylindrical or rectangular in shape, but, unlike the jet condenser, it is fitted with a large number of brass or copper tubes of small diameter (generally about ⅝ inches), through which cold water is forced by a pump called a circulating

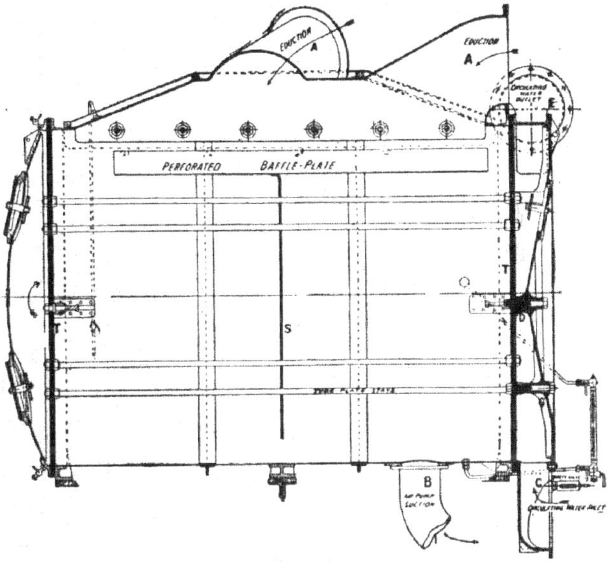

FIG. 145.

Side view of large cylindrical horizontal surface-condenser having two exhaust-inlets. The tubes are not shown. The steam enters at the orifices marked A. and is withdrawn, when condensed, through the orifice B by the air-pump. The circulating water enters at C, and is confined by the diaphragm D to the lower half of the tubes, and, having traversed these tubes, it returns through the upper half of the tubes, being finally discharged to the sea through the pipe E. T. T. are the tube-plates near the ends of the condenser casing.

pump. A vacuum is maintained in the body of the condenser by the air-pump, and the steam exhausting into this vacuum is condensed by coming in contact with the cool surface of the tubes. Or, as is often the case, the exhaust steam passes through the tubes instead of around them, and the cooling water is forced into and through the body of the condenser, the vacuum in this case being maintained in the tubes. The tubes may be placed either vertical or horizontal. When the steam is passed through the tubes, they are generally placed vertical, while, on the

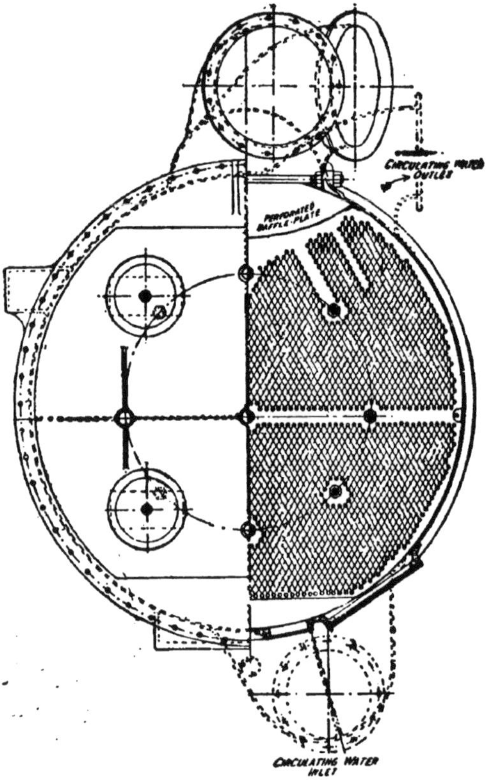

FIG. 146. END SECTIONAL VIEW OF CYLINDRICAL HORIZONTAL SURFACE CONDENSER, SHOWING A PORTION OF THE TUBES.

other hand, if the water circulates through them they are placed horizontal. The system of causing the water to circulate through the tubes, the steam surrounding them, is the more general.

Ques. 571.—How are the tubes generally arranged in a surface condenser?

Ans.—They are arranged in one or more systems, so that the condensing water passes through the condenser, usually twice, the coldest water entering at the bottom and coming in contact with the steam at its lowest temperature, and the warmest water at the top meeting the hottest steam. The exhaust steam enters at the top and after passing over the cold tubes is removed in the form of water, by the air-pump. The steam is directed in its downward course by baffle-plates, thus securing complete utilization of the cooling surface. A space is provided at the bottom of the condenser for the accumulation of the water of condensation below the cooling surface. The condenser casing or shell for naval vessels is either cast in brass or else built up from composition sheets, in order to save weight and prevent corrosion and galvanic action, which would be more liable to take place with an iron or steel shell.

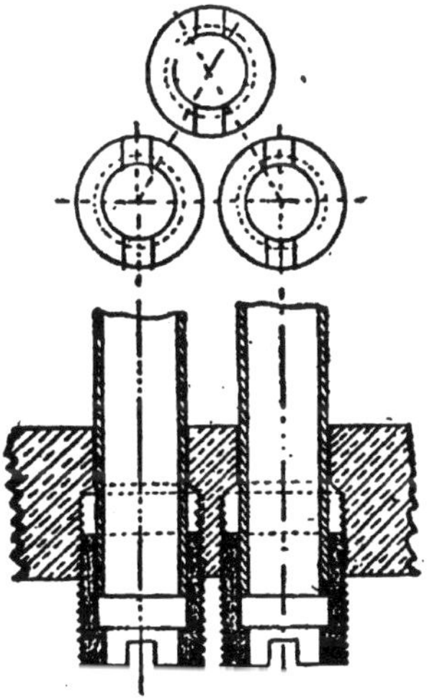

FIG. 147. DETAILS OF WICK AND GLAND PACKING FOR THE TUBES OF A SURFACE CONDENSER.

Ques. 572.—How are the tubes secured in their places?

Ans.—Brass or composition tube-plates are placed in the shell, near each end, sufficient space being left between the outside cover-plates and the tube-plates for the circulation of the cooling water. Into these plates, which are thick enough to furnish a good bearing for the tubes, the ends of the tubes are fitted and packed thoroughly tight, sometimes with a wood packing, sometimes with small screwed stuffing boxes with glands and followers, which tighten upon wick packing. The wood packing consists of a small soft wooden sleeve, which is forced into the small hole over the tube end in a dry state, and after becoming wet it swells and clamps the

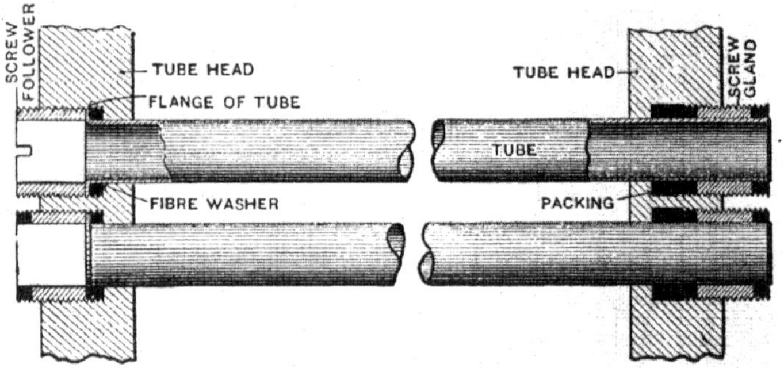

FIG. 148. METHOD OF PACKING TUBES OF A WORTHINGTON SURFACE CONDENSER.

One end of each tube is flanged and rigidly held in the tube head by means of a screw follower; the other end of the tube passes through an adjustable gland, which permits of free movement of the tube during expansion and contraction. This method of securing rigidly one end of the tube reduces the number of glands or stuffing-boxes to just one-half the number found in ordinary condensers. The glands can be readily removed and the packing replaced if it becomes leaky from long use.

tube, thus forming and preserving a tight joint so long as it is kept wet.

Ques. 573.—Which kind of packing is the most reliable for condenser tubes?

Ans.—The gland and wick, for the reason that it always remains tight, while on the other hand the wood packing will shrink and become loose if the condenser is out of service for a time.

Ques. 574.—What are the usual dimensions of the tubes of surface condensers?

Ans.—They are generally about 5/8 inches in diameter, are made of brass, about 1/20 of an inch thick, of a composition consisting of not less than 70 per cent of copper and not less than 1 per cent of tin, the remainder being zinc, the small quantity of tin being added to prevent galvanic action. The tubes are pitched not less than 31/32

Fig. 149. Worthington Surface Condenser, with Air and Circulating Pump.

inches apart in order to allow sufficient material for the gland. They are zigzagged so as to occupy as small a volume as possible. Condenser tubes vary considerably in length, depending upon the size of the condenser, the usual length in large condensers being from 8 to 10 feet, while in some very large condensers the tubes are 14 or 15 feet in length. The tube-plates are about 1 inch thick, in order to provide sufficient depth for the gland and packing for the tubes.

Ques. 575.—What type of air-pump is generally used?

Ans.—The vertical single-acting air-pump has been

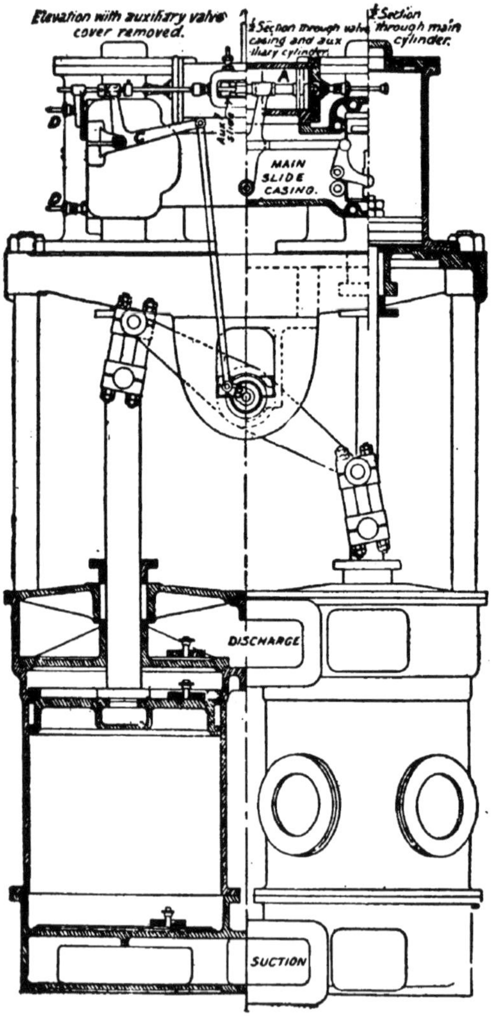

Fig. 150.

Section of Blake independent air-pump, fitted in many vessels, including several U. S. warships. There are two steam-cylinders and two single acting vertical air-pumps of the usual type. It works at slow speed and gives excellent results.

found to be the most efficient. In vertical engines the air-pump generally receives its motion from the cross-head of the engine, through the medium of a short walking-beam. There are, however, a great many engines fitted up with an independent air-pump and condenser, in which the air-pump is simply an ordinary double-acting steam-pump, having its own steam-cylinder, and may be operated independently of the engine, which is a great advantage, as there is not so much danger of the water from the condenser backing up into the cylinder in case of a sudden shut-down of the engine, which is liable to occur with a jet condenser.

Ques. 576.—Describe the parts of the vertical single-acting air-pump.

Ans.—It consists of the barrel, or cylinder, the suction-channel way at the bottom, the cover, with delivery-channel way and the hot well, the whole being made air-tight. The moving parts are the bucket, or piston, with its valves, the foot-valves and the head-valves.

Ques. 577.—Describe the arrangement of the air-pump in connection with the condenser.

Ans.—The suction-channel way is in connection with the lowest part of the condenser, in order that the water can be readily and completely removed from the condenser. It usually supports the foot-valves and all joints and valve-seat division-plates require to be fitted air-tight. The barrel is generally connected to a flange or facing of the suction-channel way, and it is constructed of composition or cast iron with a composition sleeve pressed in and bored out truly cylindrical, in order to form a smooth and durable working-cylinder for the bucket or piston, which is kept tight against the barrel, either by water-

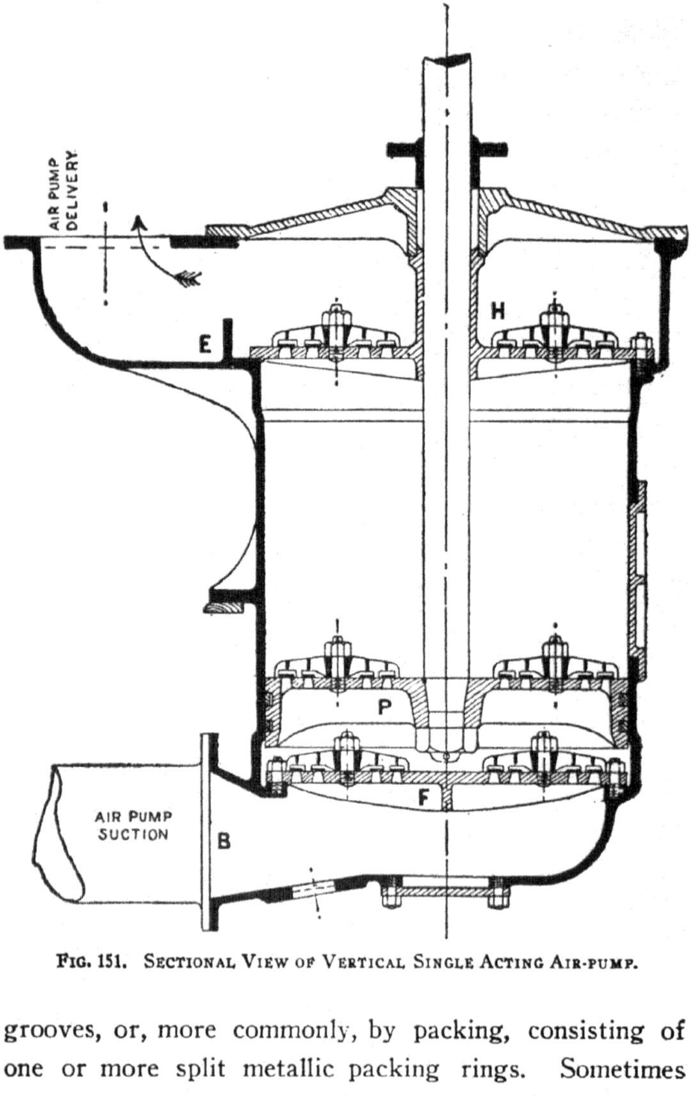

FIG. 151. SECTIONAL VIEW OF VERTICAL SINGLE ACTING AIR-PUMP.

grooves, or, more commonly, by packing, consisting of one or more split metallic packing rings. Sometimes fibrous soft packing, held in place and compressed by a follower ring, is used. A stuffing box is provided in the top cover, through which the piston-rod or trunk, as the case may be, has water-tight passage.

Ques. 578.—What kind of valves are used in air-pumps?

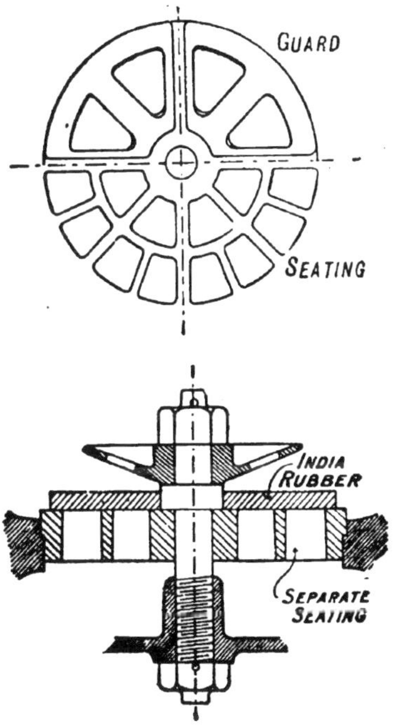

FIG. 152.
DETAILS OF RUBBER VALVE, VALVE-SEAT AND GUARD FOR AIR-PUMP.

Ans.—Rubber valves, either of hard or soft rubber, but since the introduction of mineral oil as a lubricant for the engine cylinders, it has been found that the ordinary rubber valves deteriorate under its influence, and metal valves are now largely coming into use, especially in the navies. They may be made of thin sheet metal, are light, and not affected by grease, if cleaned occasionally, and will last a long time. In form, air-pump valves are either single rectangular flaps that lift on one edge against a curved metallic guard, or else there are a number of smaller circular valves, lifting bodily from their seats, and secured to the seat by a central stud, which also carries a metal guard above the valve. The valve-seats are usually independent, being constructed of composition metal, and pressed into their places. They are divided into small spaces by gratings, so that the unsupported area of the valve may not be too large. The bucket carries the bucket-valves, which allow the air and water to pass through to the delivery side. Air-pump valves are sometimes fitted with spiral springs of bronze wire on top, to secure quick closing. The flap valves are clamped to the seat, on the stationary edge, by their curved guards.

Ques. 579.—How is the bucket or piston of the air-pump actuated?

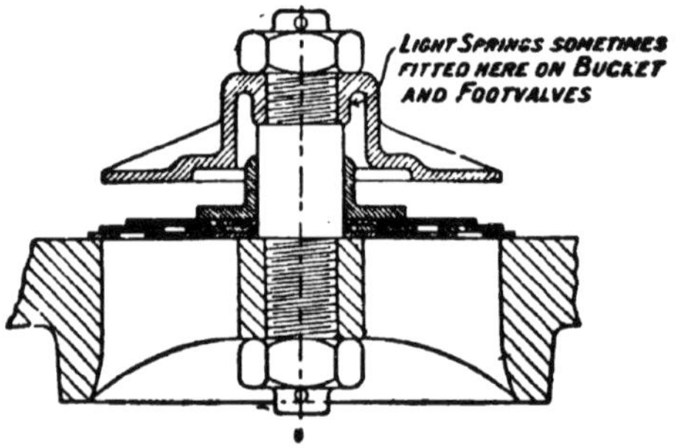

FIG. 153. SECTION OF METAL VALVE, VALVE-SEAT AND GUARD FOR AIR-PUMP.

Ans.—Either by a solid piston-rod, or by a hollow trunk, made entirely of composition, or covered by a composition sleeve. With the piston-rod type it is necessary to have a connecting rod and guides above the top cover of the air-pump, while the trunk type contains the connecting rod bearing in the trunk, near the bucket, and requires no extra guides.

Ques. 580.—What is the function of the hot well?

Ans.—It acts as a small reservoir, for the accumulation of the discharge-water from which the feed-pumps draw their supply. The later vessels in the English navy are fitted with "feed-tanks" in which the discharge from the air-pumps is allowed to accumulate, and from which the feed-pumps draw their supply of water for feeding the boilers. There is a feed-tank for each engine-room,

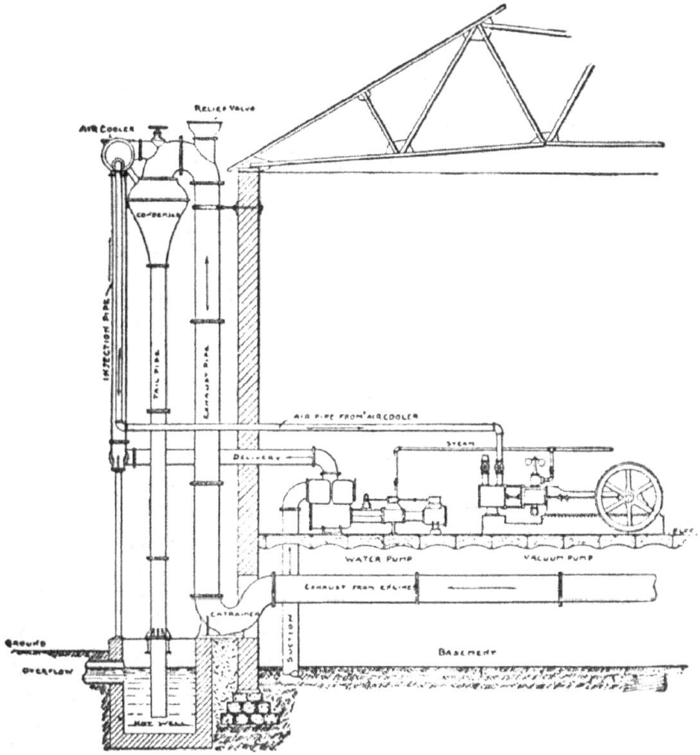

Fig. 154. Worthington Central Condenser for a Large Stationary Plant, Showing Pumps and Piping.

and they are connected by a pipe running between the two engine-rooms, fitted with a shut-off valve worked from either engine-room. These feed-tanks are fitted with glass water-gauges and zinc slabs.

Ques. 581.—What is the function of the circulating pump in connection with the surface condenser?

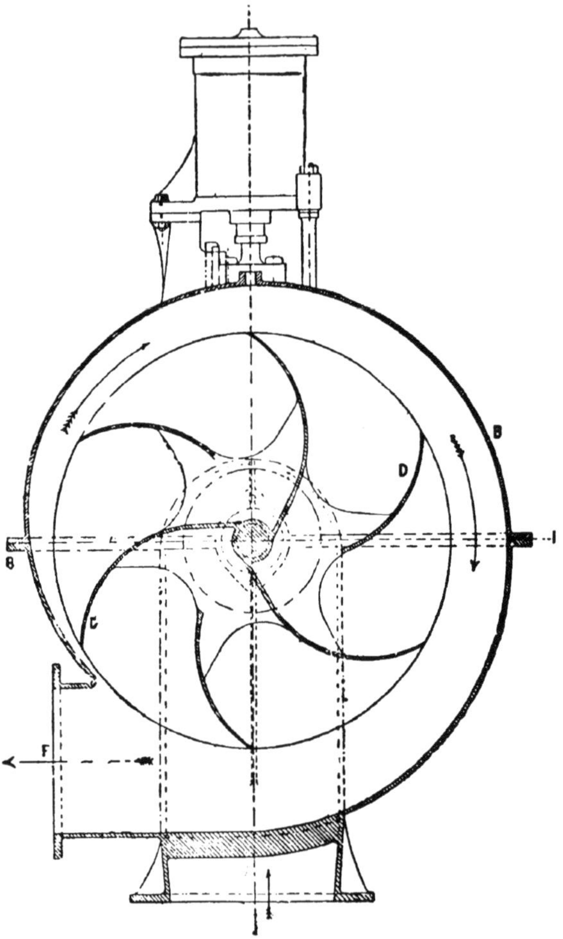

FIG. 155. TRANSVERSE SECTION OF A CENTRIFUGAL PUMP. B, CASING. D D, CURVED VANES.

Ans.—It either forces or draws the cooling water through the tubes or the body of the condenser.

Ques. 582.—What type of pump has been found to be best adapted to this work?

Ans.—The centrifugal pump worked by an independent auxiliary engine, for the reason that the pump works smoothly, there are no valves, and having a separate engine, it can be kept working and the condensers kept cool when the main engines are stopped, which is not the case with a pump that receives its motion from the main engines. Another great advantage possessed by the independent system is, that the speed may be regulated so as to supply the required quantity of water.

Ques. 583.—Describe. the construction and action of the centrifugal circulating pump.

Ans.—The pump consists of an impeller wheel or fan revolving inside a casing. The impeller and casing are made of gun metal, and the spindle or shaft carrying the impeller is either cast of gun metal in one piece with the impeller, or formed separately of forged bronze and keyed to it. This spindle runs in lignum-vitae bearings and is lubricated with water. The impeller generally consists of a central web guiding the incoming water, with two side-plates that gradually approach each other as they near the circumference and between which runs a series of curved vanes. These vanes are curved away from the direction of rotation as they proceed from the boss to the circumference. The water enters the central part of the impeller through the inlet pipe and is thrown by the rapidly revolving vanes outwards and around into the casing which surrounds the circumference of the wheel. The casing is of gradually increasing area and leads to the delivery pipe, through which it is forced by the centrifugal action to the condenser, where, after traversing the tubes, it is discharged overboard. The casing is

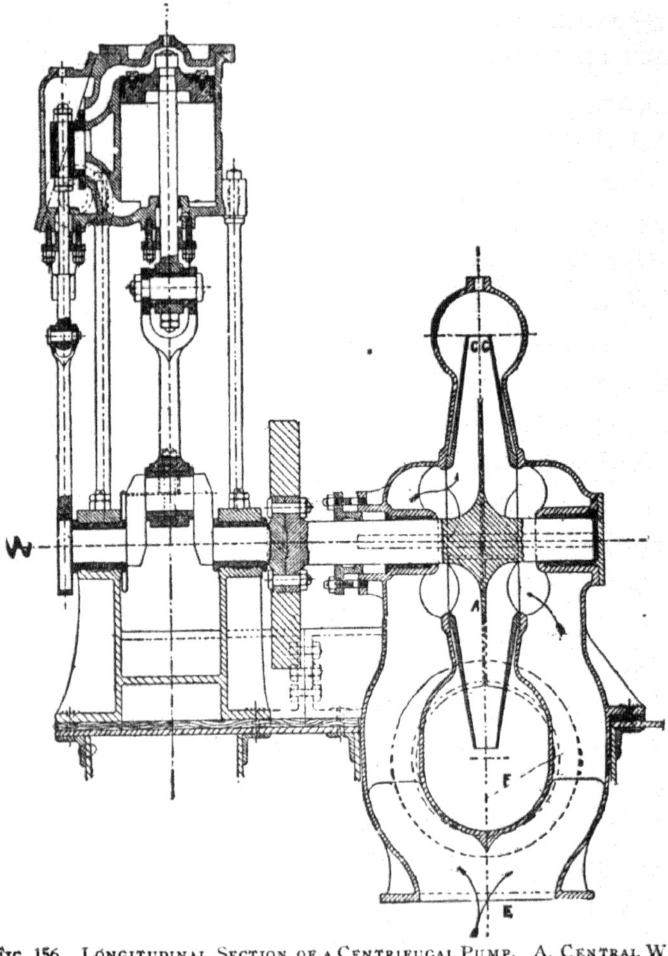

FIG. 156. LONGITUDINAL SECTION OF A CENTRIFUGAL PUMP. A, CENTRAL WEB C C, SIDE PLATES. E, INLET. F, DISCHARGE.

formed in two parts to enable the impeller to be inserted and also to facilitate inspection.

Ques. 584.—How is the quantity of water required to condense the exhaust steam of an engine determined?

Ans.—The quantity of cooling water required for a condensing system depends primarily upon the system, whether it is surface condensing or whether the condenser is a jet condenser The surface condenser needs a greater quantity of water than does the jet condenser. This is due to the fact that in the surface condenser the water, not being mixed with the steam, can not absorb the heat so rapidly.

Ques. 585.—About how much more water does a surface condenser require than is needed by a jet condenser?

Ans.—About 15 per cent more.

Ques. 586.—What three factors determine the quantity of cooling water required?

Ans.—First, the density, temperature, and volume of the steam to be condensed in a given time; second, the temperature of the overflow and third, the temperature of the injection water. For instance, it may be desired to keep the overflow at as high a temperature as possible, for the purpose of feeding the boilers, or the temperature of the injection or cooling water varies greatly. It may be 35 degrees in the winter and 70 degrees in the summer. In the marine service the temperature of sea-water varies considerably, depending upon the locality, in the tropics the temperature of the sea-water in the summer being often as high as 85 degrees Fahrenheit.

Ques. 587.—What quantity of condensing water would be required in a jet condenser into which the exhaust steam under an absolute pressure of 7 pounds is passing, assuming the temperature of the cooling water to be 55 degrees and the temperature of the overflow to be 110 degrees?

Ans.—In these calculations the total heat in the steam must be considered. This means not only the sensible heat, but the latent heat also. Now in 1 pound weight of steam at 7 pounds absolute pressure the total heat is 1,135.9 heat units. The temperature of the overflow being 110 degrees, the total heat to be absorbed from each pound weight of steam in this case would be 1,135.9 — 110 = 1025.9 thermal units. The temperature of the condensing water being 55 degrees and the temperature of the overflow being 110 degrees, there will be 110 degrees—55 degrees=55 degrees of heat absorbed by each pound of cooling water passing into and through the condenser, and the number of pounds of water required to condense each pound weight of steam under these conditions will equal the number of times 55 is contained in 1,025.9, thus, $\frac{1025}{55}$= 18.65 pounds. Assuming the steam consumption of the engine to be 17 pounds per indicated horse-power per hour, then 17 × 18.65 = 317.05 pounds of water is required per horse-power per hour for condensing purposes.

Ques. 588.—How is the weight of cooling water required per hour determined, when the steam consumption per indicated horse-power per hour is not known?

Ans.—In this case the volume of steam exhausted per hour must be considered. Thus, assume the cylinder from which the steam is exhausted to be 24 × 48 inches and the revolutions per minute to be 80. The piston displacement will equal area of piston less one-half area of rod, multiplied by length of stroke. The area of a circle 24 inches in diameter = 452.39 square inches. Suppose the piston-rod to be 4.5 inches in diameter, its area is 15.904 square inches, one-half of which = 7.952 square

Table No. 9

Jet Condensing

Quantity of Injection Water per Revolution of Engine.

INJECTION WATER 50° OVERFLOW 110°

Low-pressure Cylinder.	Single-cylinder, Water per Rev.		Two-cylinder, Water per Rev.		Three-cylinder, Water per Rev.	
	Lbs.	Galls.	Lbs.	Galls.	Lbs.	Galls.
20x36 inches..........	4.2	.5	3.9	.47	3.6	.43
22x36 "	5.1	.61	4.8	.57	4.4	.53
24x42 "	7.	.84	6.6	.79	6.	.72
26x42 "	8.3	1.	7.8	.93	7.2	.87
28x48 "	11.	1.45	10.4	1.24	9.5	1.14
30x48 "	12.6	1.52	11.7	1.41	10.8	1.3
32x54 "	16.2	1.95	15.	1.81	13.9	1.68
34x54 "	18.3	2.2	17.0	2.05	15.8	1.9
36x60 "	22.8	2.75	21.2	2.55	19.6	2.36
38x60 "	25.5	3.07	23.7	2.85	21.9	2.64
40x66 "	31.	3.73	28.8	3.45	26.7	3.2
44x66 "	37.5	4.51	34.8	4.2	32.2	3.8
48x72 "	48.5	5.84	45.	5.42	41.7	5.
52x72 "	57.	6.89	53.1	6.4	49.2	5.9
56x72 "	66.	7.9	61.5	7.41	57.	6.8
60x72 "	75.6	9.	70.5	8.5	65.3	7.8
64x72 "	85.	10.	80.	9.6	74.	8.9

(Table No. 9.—From Book on Compound Engines. By James Tribe, Detroit, Mich.)

inches. The effective area of the piston is therefore 452.39 — 7.952 = 444.4 square inches and the piston displacement equals 444.4 × 48 = 21,332.64 cubic inches. It is necessary in this calculation to express the total volume of steam exhausted per minute in cubic feet, therefore 21,332.64 ÷ 1,728 (number of cubic inches in a cubic foot) gives 12.34 cubic feet of piston displacement, and the engine running at a speed of 80 revolutions per minute will send into the condenser a volume of steam equal to twice the piston displacement multiplied by the number of revolutions per minute, expressed thus: 12.34 × 2 × 80 = 1,974.4 cubic feet per minute. Assuming the absolute pressure of the exhaust to be 7 pounds per square inch, the weight of 1 cubic foot of steam at 7 pounds absolute is .0189 pounds and the total weight of steam exhausted per minute would be 1,974.4 × .0189 = 37.3 pounds, and if 18.65 pounds of water is required to condense 1 pound weight of steam at 7 pounds absolute, the total weight of water required per minute in this case would be expressed as follows: 37.3 × 18.65 = 695.8 pounds, or per hour 695.8 × 60 = 41,748 pounds, equal to 5,029 gallons.

Ques. 589.—What quantity of condensing water would be required in a surface condenser, assuming the conditions to be the same as described in the answer to question 587?

Ans.—A surface condenser requires about 15 to 20 per cent more condensing water than a jet condenser does. It was seen in the answer referred to that 18.65 pounds of water were required to condense 1 pound weight of steam, therefore the quantity of water required by the surface condenser would be about 22 or 23 pounds for each pound of steam.

Ques. 590.—What provision is made on board of vessels for obtaining a supply of water for the condensers and for other purposes?

Table 10

Areas and Circumferences of Circles.

Diam.	Area.	Circum.	Diam.	Area.	Circum.	Diam	Area.	Circum.
.25	.049	.7854	15.5	188.692	48.694	31	754.769	97.389
.5	.1963	1.5708	16	201.062	50.265	31.25	766.992	98.175
1.0	.7854	3.1416	16.25	207.394	51.051	31.5	799.313	98.968
1.25	1.2271	3.9270	16.5	213.825	51.836	32	804.249	100.53
1.5	1.7671	4.7124	17	226.980	53.407	32.25	816.86	101.31
2	3.1416	6.2832	17.25	233.705	54.192	33	855.30	103.67
2.25	3.9760	7.0686	17.5	240.520	54.978	33.25	868.30	104.45
2.5	4.9087	7.8540	18	254.469	56.548	33.5	881.41	105.24
3	7.0686	9.4248	18.25	261.587	57.334	34	907.92	106.81
3.25	8.2957	10.210	18.5	268.803	58.119	34.25	921.32	107.60
3.5	9.6211	10.995	19	283.529	59.690	34.5	934.82	108.38
4	12.566	12.566	19.25	291.039	60.475	35	962.11	106.95
4.25	14.186	13.351	19.5	298.648	61.261	35.25	975.90	110.74
4.5	15.904	14.137	20	314.160	62.832	35.5	989.80	111.52
5	19.635	15.708	20.25	322.063	63.617	36	1017.8	113.09
5.25	21.647	16.493	20.5	330.064	64.402	36.25	1032.06	113.88
5.5	23.758	17.278	21	346.361	65.973	36.5	1046.35	114.66
6	28.274	18.849	21.25	354.657	66.759	37	1075.21	116.23
6.25	30.679	19.635	21.5	363.051	67.544	37.25	1089.79	117.01
6.5	33.183	20.420	22	380.133	69.115	37.5	1104.46	117.81
7	38.484	21.991	22.25	388.822	69.900	38	1134.11	119.38
7.25	41.282	22.776	22.5	397.608	70.686	38.25	1149.08	120.16
7.5	44.178	23.562	23	415.476	72.256	38.5	1164.15	120.95
8	50.265	25.132	23.25	424.557	73.042	39	1194.59	122.52
8.25	53.456	25.918	23.5	433.731	73.827	39.25	1209.95	123.30
8.5	56.745	26.703	24	452.390	75.398	39.5	1225.42	124.09
9	63.617	28.274	24.25	461.864	76.183	40	1256.64	125.66
9.25	67.200	29.059	24.5	471.436	76.969	40.25	1272.39	126.44
9.5	70.882	29.845	25	490.875	78.540	40.5	1288.25	127.23
10	78.540	31.416	25.25	500.741	79.325	41	1320.25	128.80
10.25	82.516	32.201	25.5	510.706	80.110	41.25	1336.40	129.59
10.5	86.590	32.986	26	530.930	81.681	41.5	1352.65	130.37
11	95.033	34.557	26.25	541.189	82.467	42	1385.44	131.94
11.25	99.402	35.343	26.5	551.547	83.252	42.25	1401.98	132.73
11.5	103.869	36.128	27	572.556	84.823	42.5	1418.62	133.51
12	113.097	37.699	27.25	583.208	85.608	43	1452.20	135.08
12.25	117.859	38.484	27.5	593.958	86.394	43.25	1469.13	135.87
12.5	122.718	39.270	28	615.753	87.964	43.5	1486.17	136.65
13	132.732	40.840	28.25	626.798	88.750	44	1520.53	138.23
13.25	137.886	41.626	28.5	637.941	89.535	44.25	1537.86	139.01
13.5	143.130	42.411	29	660.521	91.106	44.5	1555.28	139.80
14	153.938	43.982	29.25	671.958	91.891	45	1590.43	141.37
14.25	159.485	44.767	29.5	683.494	92.677	45.25	1608.15	142.15
14.5	165.130	45.553	30	706.860	94.248	45.5	1625.97	142.94
15	176.715	47.124	30.25	718.690	95.033	46	1661.90	144.51
15 25	182.654	47.909	30.5	730.618	95.818	46.25	1680.01	145.29

TABLE 10—*Continued.*

Diam.	Area.	Circum.	Diam.	Area.	Circum.	Diam.	Area.	Circum.
46.5	1698.23	146.08	62.25	3043.47	195.56	78	4778.37	245.04
47	1734.94	147.65	62.5	3067.96	196.35	78.25	4809.05	245.83
47.25	1753.45	148.44	63	3117.25	197.92	78.5	4839.83	246.61
47.5	1772.05	149.22	63.25	3142.04	198.71	79	4901.68	248.19
48	1809.56	150.79	63.5	3166.92	199.50	79.25	4932.75	248.97
48.25	1828.46	151.58	64	3216.99	201.06	79.5	4963.92	249.76
48.5	1847.45	152.36	64.25	3242.17	201.85	80	5026.56	251.33
49	1885.74	153.93	64.5	3267.46	202.68	80.5	5089.58	252.90
49.25	1905.03	154.72	65	3318.31	204.20	81	5153.00	254.47
49.5	1924.42	155.50	65.25	3343.88	204.99	81.5	5216.82	256.04
50	1963.50	157.08	65.5	3369.56	205.77	82	5281.02	257.61
50.25	1983.18	157.86	66	3421.20	207.34	82.5	5345.62	259.18
50.5	2002.96	158.65	66.25	3447.16	208.13	83	5410.62	260.75
51	2042.82	160.22	66.5	3473.23	208.91	83.5	5476.00	262.32
51.25	2062.90	161.00	67	3525.66	210.49	84	5541.78	263.89
51.5	2083.07	161.79	67.25	3552.01	211.27	84.5	5607.95	265.46
52	2123.72	163.36	67.5	3578.47	212.06	85	5674.51	267.04
52.25	2144.19	164.14	68	3631.68	213.63	85.5	5741.47	268.60
52.5	2164.75	164.19	68.25	3658.44	214.41	86	5808.81	270.17
53	2206.18	166.50	68.5	3685.29	215.20	86.5	5876.55	271.75
53.25	2227.05	167.29	69	3739.28	216.77	87	5944.66	273.32
53.5	2248.01	168.07	69.25	3766.43	217.55	87.5	6013.21	274.89
54	2290.22	169.64	69.5	3793.67	218.34	88	6082.13	276.46
54.25	2311.48	170.43	70	3848.46	219.91	88.5	6151.44	278.03
54.5	2332.83	171.21	70.25	3875.99	220.70	89	6221.15	279.60
55	2375.83	172.78	70.5	3903.63	221.48	89.5	6291.25	281.17
55.25	2397.48	173.57	71	3959.20	223.05	90	6371.64	282.74
55.5	2419.22	174.35	71.25	3987.13	223.84	90.5	6432.62	284.31
56	2463.01	175.92	71.5	4015.16	224.62	91	6503.89	285.88
56.25	2485.05	176.71	72	4071.51	226.19	91.5	6573.56	287.46
56.5	2507.19	177.5	72.25	4099.83	226.98	92	6647.62	289.03
57	2551.76	179.07	72.5	4128.25	227.75	92.5	6720.07	290.60
57.25	2574.19	179.85	73	4185.39	229.34	93	6792.92	292.17
57.5	2596.72	180.64	73.25	4214.11	230.12	93.5	6866.16	293.74
58	2642.08	182.21	73.5	4242.92	230.91	94	6939.79	295.31
58.25	2664.91	182.99	74	4300.85	232.48	94.5	7013.81	296.88
58.5	2687.83	183.78	74.25	4329.95	233.26	95	7088.23	298.45
59	2733.97	185.35	74.5	4359.16	234.05	95.5	7163.04	300.02
59.25	2757.19	186.14	75	4417.87	235.62	96	7238.25	301.59
59.5	2780.51	186.92	75.25	4447.37	236.40	96.5	7313.80	303.16
60	2827.44	188.49	75.5	4476.97	237.19	97	7389.81	304.73
60.25	2851.05	189.28	76	4536.37	238.76	97.5	7466.22	306.30
60.5	2874.76	190.06	76.25	4566.36	239.55	98	7542.89	307.88
61	2922.47	191.64	76.5	4596.35	240.33	98.5	7620.09	309.44
61.25	2946.47	192.42	77	4656.63	241.90	99	7697.70	311.02
61.5	2970.57	193.21	77 25	4686.92	242.69	99.5	7775.63	312.58
62	3019.07	194.78	77.5	4717.30	243.47	100	7854.00	314.16

Ans.—All holes in the hull of a ship below the water-line for the supply or discharge of condensing water, or for any other purpose, are fitted with valves having long spindles which are brought inside the vessel through stuffing boxes, in order that the valves may be worked from inboard. The circulating pumps take their suction from a large screw-down inlet valve on the bottom of the ship, while the discharge is through similar valves on the ship's side.

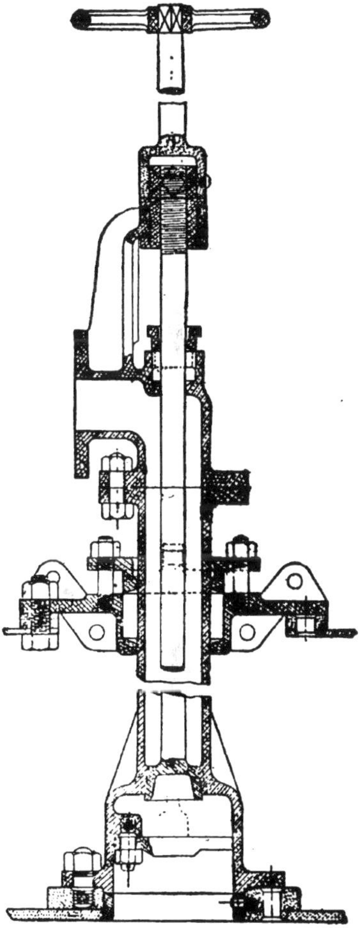

FIG. 157. SEA-VALVE.

Ques. 591.—What type of valve is largely used for this purpose?

Ans.—The Kingston sea-valve. Strainers are placed over all inlets, to prevent the entrance of weeds and other impurities.

CHAPTER VIII

AUXILIARY MACHINERY AND FITTINGS

Ques. 592.—Besides the air and circulating pumps, what other pumps are required in well-equipped steam plants, or aboard steam-ships?

Ans.—Boiler feed-pumps, fire service, pumps for hydraulic elevators, and other service requiring water-pressure, and in addition, on ship-board, pumps are required for emptying the bilges and tanks and for supplying water for washing the decks, evaporator service and for sanitary purposes.

Ques. 593.—Is there a special pump provided for each service?

Ans.—Not in all cases, but one pump may be connected in such a manner as will permit of its being used alternately for several different purposes. However, a special pump is, or at least should always be provided for feeding the boilers. Also a special bilge-pump is usually supplied, for the reason that it handles very dirty water, that should not be passed through any other pipe system. In small vessels one pump (the donkey) usually serves for nearly all purposes, including auxiliary boiler-feed, and on Western river steamers an independent pump (the doctor) having a steam-cylinder and walking-beam, drives a system of pumps for feed, fire and bilge-pumping service.

Ques. 594.—What special features should appertain to the boiler feed-pump?

Ans.—It should be simple, durable, of great strength and ample capacity to insure regular and reliable service under the most severe conditions. It is always best to have the main and auxiliary feed-pumps duplicates of each other if possible, for the reason that in cases of emergency the different parts are interchangeable. In the marine service the main feed-pump draws its supply of water from the hot well, feed-heater or the feed-tank, as the case may be. The auxiliary or duplicate feed-pump may be arranged so as to draw from either of these sources, and also from the sea, thus making provision for emergency.

FIG. 158. THE WORTHINGTON BOILER FEED PUMP, ADMIRALTY PATTERN FOR 250 POUNDS PRESSURE.

Ques. 595.—Where should the feed-pumps be located?

Ans.—As near to the boiler-room as possible, in order that the engineer in charge of the boilers may have full control of the feed-water supply. On board of vessels, when the feed-pump is worked from the main engine, the auxiliary, or injector is usually placed in the stoke-hold.

Ques. 596.—What type of boiler feed-pump has

been found to be the most reliable for all kinds of service?

Ans.—The double acting steam-pump, working independently of all other machinery. The horizontal variety is principally used for land service, while on board steam vessels the vertical type is preferred, for the reason that it occupies less floor space. In both the horizontal and vertical types, the water valve-chambers have removable covers, allowing a ready access to the valves and valve-seats. The steam-valves of these pumps are actuated in various ways. In the duplex variety, which consists of two pumps combined into one, the steam-valve of one side is moved from the piston-rod of the other, and vice versa, while with a pump having but a single steam-cylinder, the steam-valve is worked by a tappet action from its own piston-rod.

Ques. 597.—What two varieties of feed-pumps are largely in use on ocean steamers?

Ans.—The Weir vertical double-acting steam-pump and the Belleville, which is built either vertical or horizontal. In the Weir pump the water-valves are a series of small cones milled out of solid metal and give a large area of opening with a slight lift. The steam-valve arrangement of the Weir pump is rather complicated and requires to be maintained in perfect condition, to insure good service. It consists of a main valve for distributing steam to the cylinder and an auxiliary valve for distributing steam to work the main valve. The main valve moves horizontally from side to side, being driven by steam admitted and exhausted from each end alternately. The auxiliary valve is actuated by a lever with a fixed fulcrum worked by the piston-rod of the pump. This auxiliary valve moves on a flat face on the back of the main valve and in a direction at right angles to the latter. Both the main and auxiliary valves are simply slide valves, but the main valve is half round, the round side working on the correspondingly shaped cylinder port-seat, while the back of the valve is flat and forms the seat for the auxiliary valve. Both ends of the main valve are lengthened, so as to project beyond the port face and are turned cylindrical, with flat ends. Caps are fitted on each of these ends, forming cylinders which are closed at the mouths by the flat ends of the main valve, which act as pistons, the length of stroke

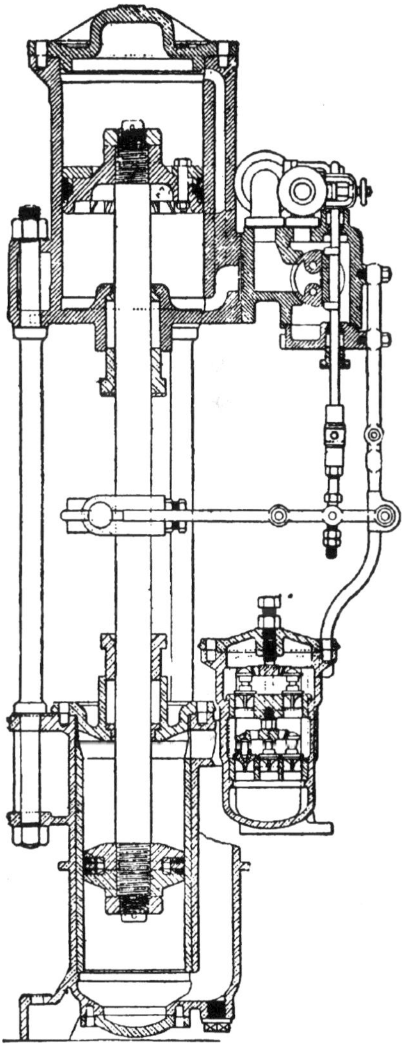

FIG. 159. WEIR'S FEED-PUMP FOR MARINE SERVICE.

that the piston can make being the full travel of the valve. The auxiliary valve-seat has three ports, the center one being the exhaust and the two side ports being steam-passages leading through the piston ends of the main valve. The right-hand cylinder port-passage is led through the left-hand end of the piston and the left-hand passage leads to the other end. These ports admit steam to the two small caps or cylinders at each end of the valve alternately, by which it is thrown from side to side. Besides the ports already referred to, there are two other ports formed on the auxiliary valve-seat leading to and corresponding to two ports on the half-round seat of the main valve. These ports are for the purpose of admitting steam to the top and bottom of the main cylinder, and are arranged on the auxiliary valve-seat to cut off steam before the end of the stroke, and so reduce the speed of the piston, but the expansion chambers at each end of the main valve are fitted with by-passes to admit steam for the full stroke when so desired. This may be necessary when starting the pump, as then the water-cylinder may be full of water. These by-passes are formed by notches cut in the edges of the caps and may be opened or closed by turning the caps by means of spindles provided at each side of the valve-chest, and thus give a definite cut-off. There are separate by-passes for the up and down strokes, and the silent working of the pump depends upon the proper adjustment of these by-passes.

Ques. 598.—Describe the action of the Belleville feed-pump.

Ans.—The pump is double-acting, having an ordinary

flat slide-valve without lap, worked by a curved lever, which is moved at each end of the stroke, by a projection or lug on the piston-rod. The steam-ports are arranged at each end of the cylinder in such a manner as to admit the steam uniformly all around the cylinder circumference, and not at the top only, which prevents bending forces on the rod. The steam-pressure remains constant, therefore, until near the end of the stroke, when the projection strikes the valve-lever and commences to close the

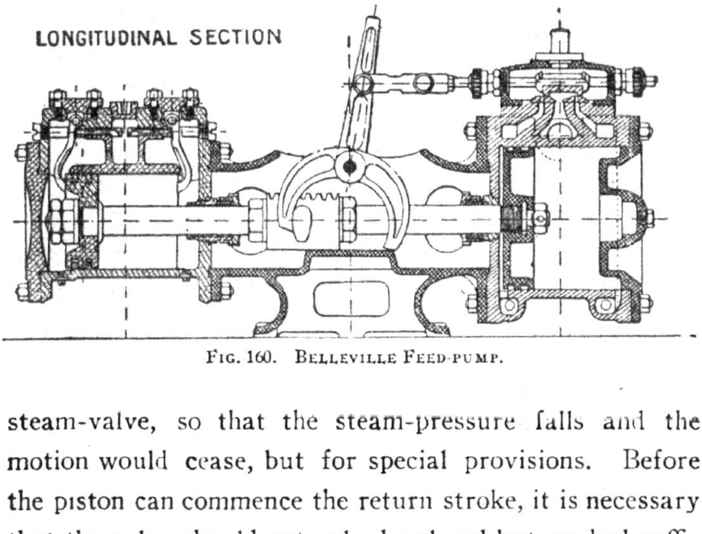

FIG. 160. BELLEVILLE FEED-PUMP.

steam-valve, so that the steam-pressure falls and the motion would cease, but for special provisions. Before the piston can commence the return stroke, it is necessary that the valve should not only be closed but pushed sufficiently far over to reopen for steam on the other side. To enable the steam already in the cylinder to complete the stroke and throw the valve over to the opposite side, an orifice is provided at each end of the water-cylinder, closed by levers and communicating with the suction-chamber, so that when the water-piston nears the end of its stroke, it strikes one of these levers and opens the orifice to the suction-chamber, thus causing the pressure in the water-cylinder to fall, and the steam, although cut off, is enabled by its expansive force to push the piston to the end of the stroke and reverse the valve. When motion begins in the opposite direction the water-valves are a series of small valves, generally eight in number, at each end, four for suction and four for discharge. Small holes about $\frac{1}{16}$ inch in diameter, are made through the levers into

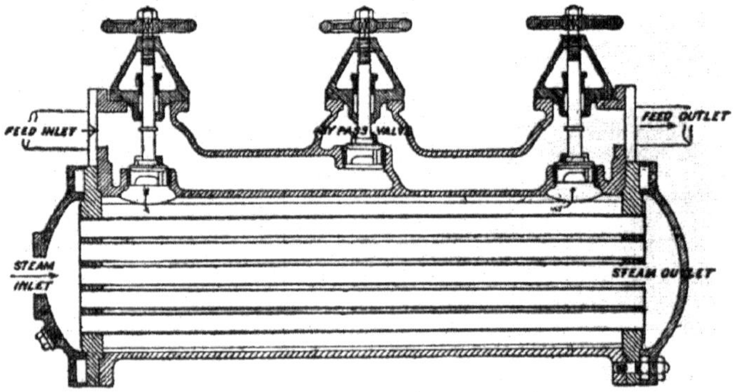

FIG. 161. KIRKALDY'S FEED-HEATER.

the passage leading to the suction chamber, so that a small quantity of water is always escaping from the water-cylinder, which causes the pump to keep slowly in motion, even when the feed-valves on the boilers are closed.

Ques. 599.—Are feed-water heaters much in use in the marine service?

Ans.—They are largely used in the mercantile service, and results justify their adoption.

Ques. 600.—Describe the construction and operation of Kirkaldy's feed-heater.

Ans.—It is constructed along lines similiar to a surface condenser, having tubes rolled into tube-plates in the ordinary manner, the whole surrounded by an outside shell, leaving spaces at each end between the tube-plates and end-covers. The feed-water does not mix with the heating steam, but is drawn through the tubes, on the

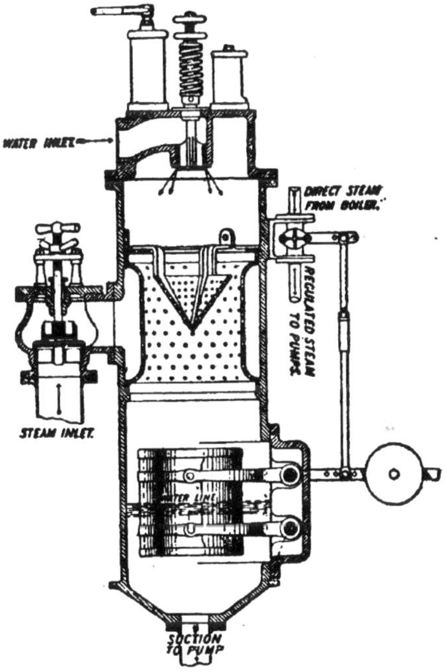

FIG. 162. WEIR'S FEED-HEATER AND REGULATOR.

outside of which is the steam, which is usually the exhaust from various auxiliary engines, or it may be drawn from the boilers. By-pass valves are fitted, so that when necessary the feed-water can be passed direct, without passing through the heater.

Ques. 601.—Describe the construction and operation of Weir's feed-heater and regulator.

Ans.—It takes steam from the final receiver of the engine after it has done most of its work. The steam enters the heating chamber through a circular perforated ring and there mixes with the cold feed-water, which is admitted through the spring-loaded valve on the cover.

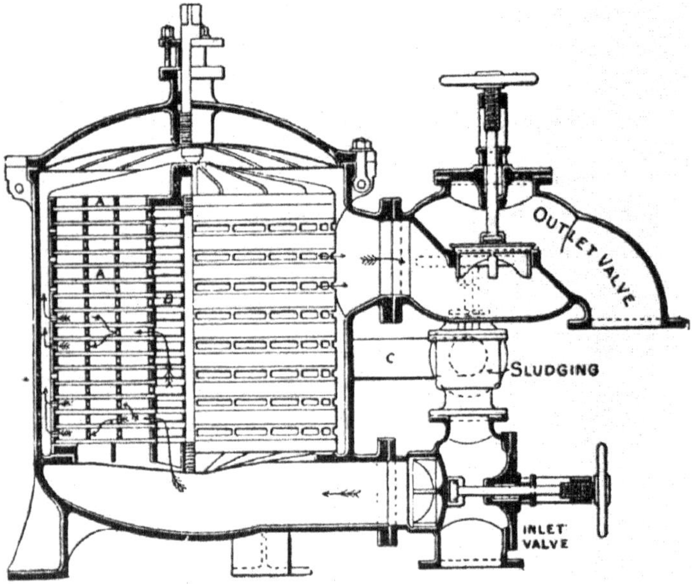

Fig. 163. The Harris Grease Filter.

The heated water falls to the bottom of the heater, from whence it is removed by the feed-pump. A galvanized iron float is fitted to the bottom of the heater, which communicates by means of levers with the steam-valve leading to the feed-pump, thus keeping the water-level constant in the heater and preventing the pumps from drawing air.

Ques. 602.—What provision is made on board steam-vessels for the prevention of oil or grease passing into the boilers along with the feed-water?

Ans.—Numerous types of grease-filters are in use. In the Harris grease-filter the feed-water is caused to pass through a series of gratings, on each of which is fitted one or two sheets of filtering material, consisting of toweling or flannel, supported by wire gauze. When the cloths become dirty they are cleaned by a steam jet, and washed off by a reverse current of water.

Ques. 603.—What is the object of placing a governor on an engine?

Ans.—To maintain regularity of speed of the engine when the load is varied from any cause.

Ques. 604.—Upon what principle do the most of the governors for land engines operate?

Ans.—Upon the principle of centrifugal force causing two balls or weights, each suspended or attached to a lever swinging on a fulcrum, fixed near the top of a vertical revolving spindle, to fly outward as the speed increases; and the force of gravitation which acts in the opposite direction as the speed decreases. The outward movement of the balls or weights is utilized to either close the throttle or shorten the point of cut-off, while the inward movement has the opposite effect.

Ques. 605.—Are governors required on marine engines?

Ans.—They are, for the reason that in a marine engine considerable diminution in resistance may ensue in rough or stormy weather, from the pitching motion of the vessel, which causes the propellers to rise partly out of the water, thus causing what is technically known as "racing of the engines."

Ques. 606.—Is the centrifugal type of governor suitable for marine service?

Ans.—It is not, for the reason that the forces acting upon the balls or weights would be affected by the motion of the ship and the action would be irregular. Other forms of governors for marine engines are in use with various degrees of success, but all, or nearly all of them, possess the one defect of requiring an increased speed of the engine to cause them to act, and even then their action is sluggish, the throttle-valve being generally closed after the racing is over.

Ques. 607.—What type of marine governor is likely to prove the most successful in marine service for the prevention of "racing?"

Ans.—A governor that acts by variations of pressure at the stern of the vessel near the propeller, and not from engine-speed variations. Racing being caused by diminished immersion of the propeller, it is accompanied by a diminution of pressure of water at that part, which can be utilized to actuate the throttle-valve. Such governors may therefore anticipate and prevent any increase of speed due to the above cause, although they would have no effect in case of a serious increase of speed, due to such an accident as a broken shaft or propeller.

Ques. 608.—Describe Dunlop's governor, which is of the latter type.

Ans.—It consists of a sea-cock at the stern of the

ship, opening into an air-vessel or air-chamber, so constructed that, by opening the sea-cock, water flows into the air-vessel and compresses the air contained therein to a pressure equivalent to the head of water outside the ship. From the top of the air-chamber a pipe is led to the under side of an air-tight elastic diaphragm, forming part of an apparatus in the engine-room. On the upper side of the diaphragm is a spiral spring, with means of adjusting its compression to balance the air pressure below the diaphragm. From the center of the diaphragm a connection is made to the slide-valve of a small steam-cylinder so constructed that its piston moves in exact accordance with the movements of the diaphragm. This steam-piston is connected by suitable gear to the throttle-valve of the engine whose speed is to be controlled. The action is as follows: The sea-cock being open, any variation of head of water outside the ship is accompanied by an inflow or outflow of water through it and consequently a variation in the pressure of the air contained in the air-chamber, and also under the diaphragm of the engine-room apparatus, causing the diaphragm to move through such part of its travel as is requisite to enable the compression of spring and the air-pressure to balance each other again. Every movement of the diaphragm is followed by a corresponding movement of the governor steam-piston, and consequently of the throttle-valve of the engines under control, the time taken between the variation in the head of water at the stern of the ship and the moving of the throttle-valve being practically nothing. The governor therefore anticipates any increase in the speed of the engines due to the propeller rising out of the water and does not depend upon a variation in speed of the engines to be controlled, before it acts. By adjusting the balance between the spring and the air-pressure under the diaphragm the diaphragm begins to fall and the throttle-valve to close, when the tips of the propeller-blades rise

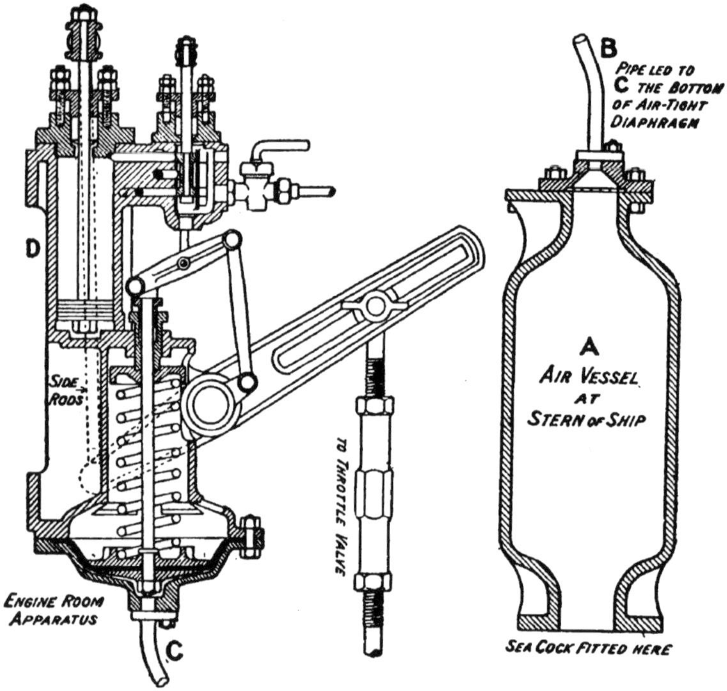

FIG. 164. DUNLOP'S GOVERNOR.

to any desired distance above the surface of the water. The air-vessel should be fitted as far aft in the screw-tunnel as possible, the hole through the side of the vessel being placed about one-fourth the diameter of the propeller below the level of the center of the shaft. The reports of the action of this governor in the mercantile marine are very satisfactory. It is fitted in the "Campania," "Paris," and many other vessels.

Ques. 609.—How is the fresh water needed on board ship for drinking, washing, culinary purposes, and for making up for the waste of feed-water for the boilers and for various other purposes, obtained?

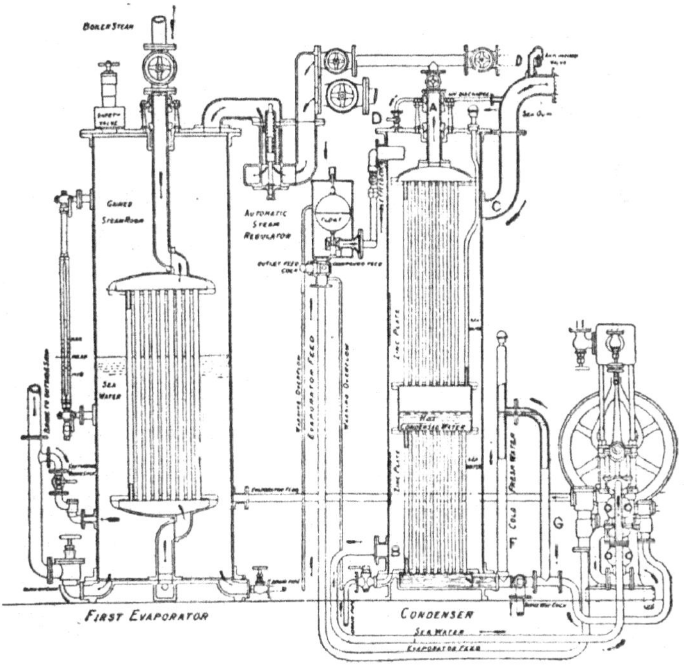

NORMANDY'S EVAPORATOR.

Ans.—By means of evaporators and distillers. The evaporators are really small boilers, with heat obtained from steam passing through tubes, while the water to be evaporated surrounds the tubes. There is no coal used in these boilers, the steam being obtained from the main boilers. The vapor produced is conducted to the distilling apparatus, where it is condensed into fresh drinking water, and a portion of it goes to the condensers for the purpose of making up the deficiency of boiler feed-water. The condensed primary steam is returned to the boilers.

Ques. 610.—Describe Normandy's evaporator.

Ans.—In this type of evaporator the tubes are all straight and rolled into tube-plates at their ends. The steam from the main boilers enters these tubes through a pipe at the top and evaporates the surrounding sea-water contained in the shell, and is itself condensed and passes out through the bottom, returning to the boilers. The vapor generated outside the tubes is conveyed by a valve and pipe, either to the auxiliary condenser for feed-water make-up, or else to the distilling condensers for the production of drinking water. The resulting scale is deposited in the evaporator, from whence it is cleaned at intervals. The sea-water for the evaporator is supplied by a pump. It takes its supply from a feed-box containing a float which maintains a constant level in the feed-box.

Ques. 611.—Describe Normandy's condenser.

Ans.—The steam from the evaporator enters the condenser through a pipe at the top and passes downwards through two series of tubes, the upper set being the condensing and the lower the cooling tubes. These tubes are surrounded by a casing, which is kept filled with cold sea-water that enters at the bottom and flows out at the top through an overflow pipe that is connected to the casing at a point a short distance below the top and is then carried to some distance above the top of the chamber before discharging overboard. By means of this arrangement the hottest sea-water is not discharged overboard, but instead may be used in the evaporator, in connection with the condenser, and thus promote economy of evaporation. An air-pipe is fitted to allow the air evolved from the condensing water in the casing by heat to pass into the overflow pipe leading to the sea. The condensed water rises from the lower chamber through a stand-pipe connected at the bottom and overflows from this pipe into and down another pipe leading to the suction of a small steam donkey pump, which pumps it into test-tanks, from whence it flows by gravity to the water-tanks in the hold of the vessel. By this arrangement the cooling tubes of the condenser are always kept full of water and the fresh water is drawn off cold.

Ques. 612.—On vessels carrying cargoes of fresh meat and other perishable articles that are affected by the heat, what provision is made for their preservation?

Ans.—Various types of refrigerating machinery are in use, some using the cold-air system, others the carbonic-acid system, and a few of the smaller ships are fitted with machines for making ice only.

Ques. 613.—Describe the cold-air system.

Ans.—The machine consists of a tandem compound engine having piston slide-valves both on the same valve-rod and worked by a single eccentric. This engine supplies the motive power of the apparatus. Two air-cylinders, one called the compressing cylinder and the other one the expanding cylinder, are placed side by side and in line with the low-pressure cylinder of the engine. These air-cylinders are double acting, the pistons receiving their motion from the crank-shaft driven by the engine. The action of the device is simple and is as follows: The revolving shaft, through the medium of connecting rods and guides, moves the pistons up and down. Air is drawn into the compressing cylinder through inlet-valves from the surrounding atmosphere or from the cold room. It is compressed on the return stroke of the piston and passes into the cooling chamber, which is constructed similar to a surface condenser, having a pump to circulate the cooling sea-water through it. The work done thus far appears as heat in the air and this heated air, passing through the tubes of the air-cooler, is cooled by the circulating water and is then led to the valve-chamber of the expanding cylinder. The valve arrangement of this cylinder consists of a slide-valve and an expansion valve working on the back of the slide-valve. This arrangement supplies a means of sharply cutting off the inlet of air when it enters the expanding cylinder. The compressing cylinder is provided with a water-jacket through which the circulating pump delivers the cooling water on its way from the air-cooler to the sea. The slide-valves are so arranged in the expanding cylinder that when the proper quantity of air is admitted the supply is cut off and during the remainder of the stroke the air expands and therefore does work on the piston and heat is expended in the process in exactly the converse manner to the generation of heat in the compressing cylinder. As, however, the air has been deprived of its surplus heat

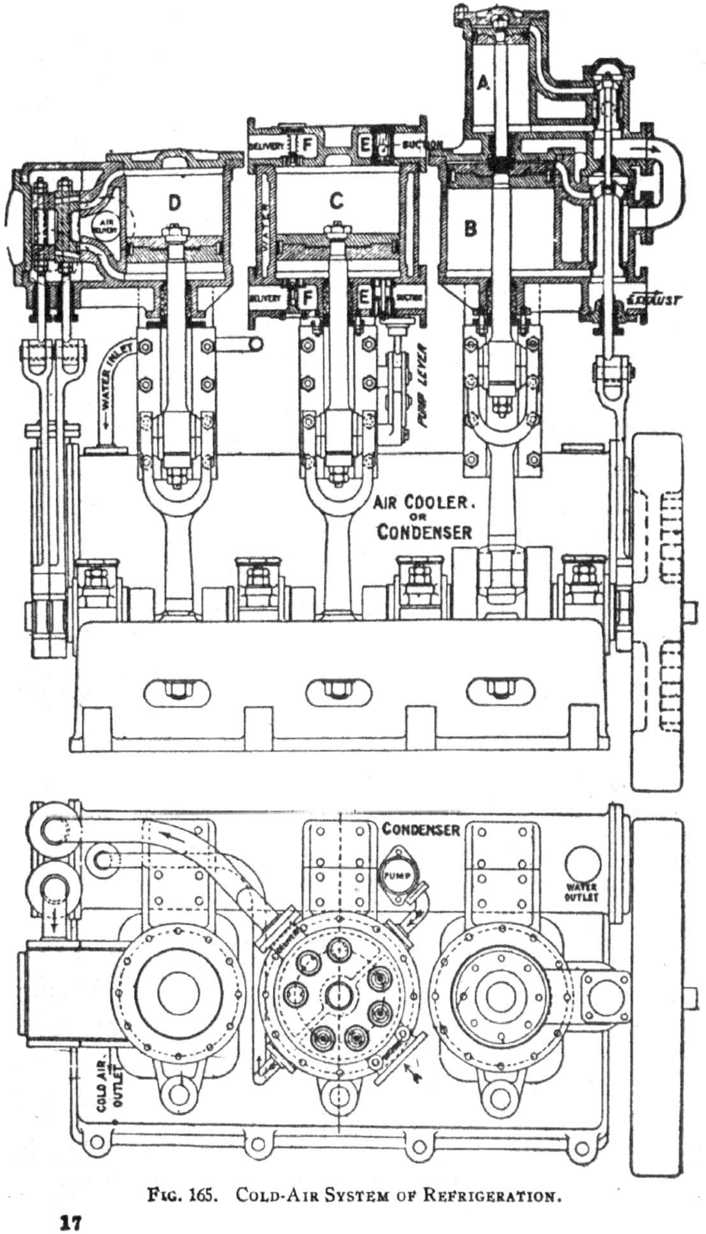

FIG. 165. COLD-AIR SYSTEM OF REFRIGERATION.

in the cooling chamber, the heat equivalent of the work it does in the expanding cylinder is absorbed from itself and the result is a considerable lowering of its temperature. This cold air is then exhausted through the orifice of the slide-valve in the usual manner, and conducted first to the

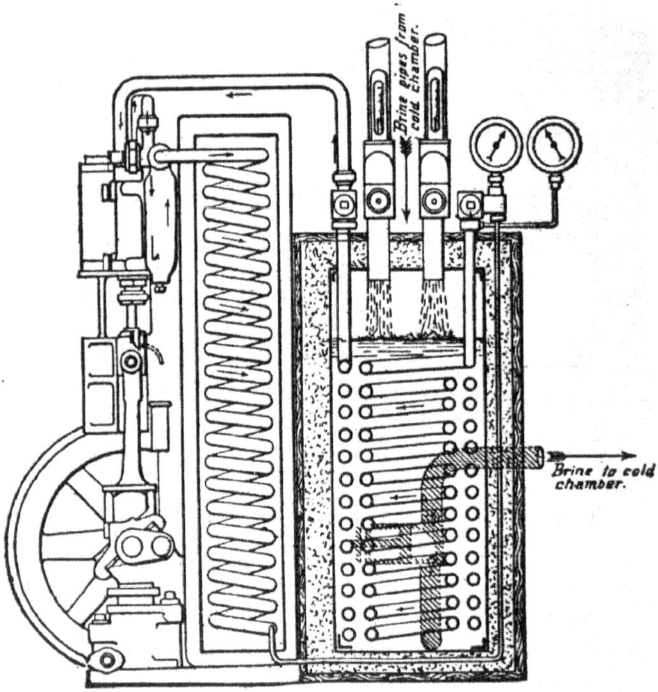

FIG. 166. CARBONIC-ACID SYSTEM OF REFRIGERATION.

"snow-box" a small accessible chamber in which the snow formed from the moisture is deposited, and from thence to the cold chamber, in which the supply of meat or provisions is kept and where it displaces air of a higher temperature. The refrigerating chamber is insulated by lagging its bulkheads, ceiling, and floor with silicate cotton or other non-conductor, a teak lining being fitted over this to form the inside surface.

Ques. 614.—Describe the carbonic-acid system.

Ans.—A very successful and efficient device is the carbonic-anhydride system of Messrs. J. & E. Hall, in which carbonic anhydride is passed round continually in the circuit. The apparatus consists of three parts: a compressor, a condenser, and an evaporator. The compressor draws in heated and expanded gas from the evaporator and compresses it. The compressed gas then passes to a condenser, consisting of coils in which the warm compressed gas is cooled and liquefied by reduction of temperature caused by the action of the cooling sea-water. From the condenser the cool liquid carbonic anhydride is conveyed into the evaporator consisting of coils, where it vaporizes and expands, absorbing heat in the process and cooling the surrounding brine, which is in contact with the coils. This cold brine is circulated by a small pump to the refrigerating chamber, where it is conducted through a long series of rows of cooling pipes, termed "grids," which are placed at the roof of the chamber. The cold-brine "grids" in this position set up a circulation of air, the cold air descending and being replaced by air not so cold, which is cooled in its turn. Any moisture in the air is condensed on the "grids" and appears as frost on the pipes. The theory of the action of this system is as follows: Under atmospheric pressure the liquid CO_2 would evaporate at a temperature of 120 degrees Fahrenheit below zero, but its temperature of evaporation rises with the pressure, in a similar manner as water. At a pressure of 500 pounds per square inch it boils at a temperature of 30 degrees Fahrenheit so that cold water may be used to supply the heat for boiling it. The pressure in the evaporator is therefore regulated to the required temperature of the cooling water, so that a considerable pressure is necessary in the evaporator. The compressor draws the gas from the evaporator and compresses it to the liquefying pressure, the heat due to the compression being absorbed by the cooling water in the condenser coils and the gas in these coils becomes liquid before its exit. The liquid is then boiled in the evaporator coils, cooling the surrounding brine by the heat absorbed during evaporation. The compressor gland is made tight by cupped leathers with glycerine forced between them at a higher pressure than that in the compressor, so that no escape of gas can take place. The carbonic anhydride is supplied in steel cylinders to replenish the supply.

Ques. 615.—What types of dynamos are used on board ships for generating electric current for internal illumination and for working search-lights and motors?

Ans.—They are usually of the two-pole type, direct driven and carried on an extension of the engine-bed. They have drum armatures and the field-magnets are compound wound, to give a constant pressure of 80 or 100 volts for any current from zero to the maximum, while the speed is maintained constant. The usual speed is 320 revolutions per minute. The machines are connected to a switchboard located in a central position, from which the current is distributed to the various circuits for lighting, motors, etc. This board is so arranged that a circuit can be quickly changed from one machine to another, but no circuit can receive current from two

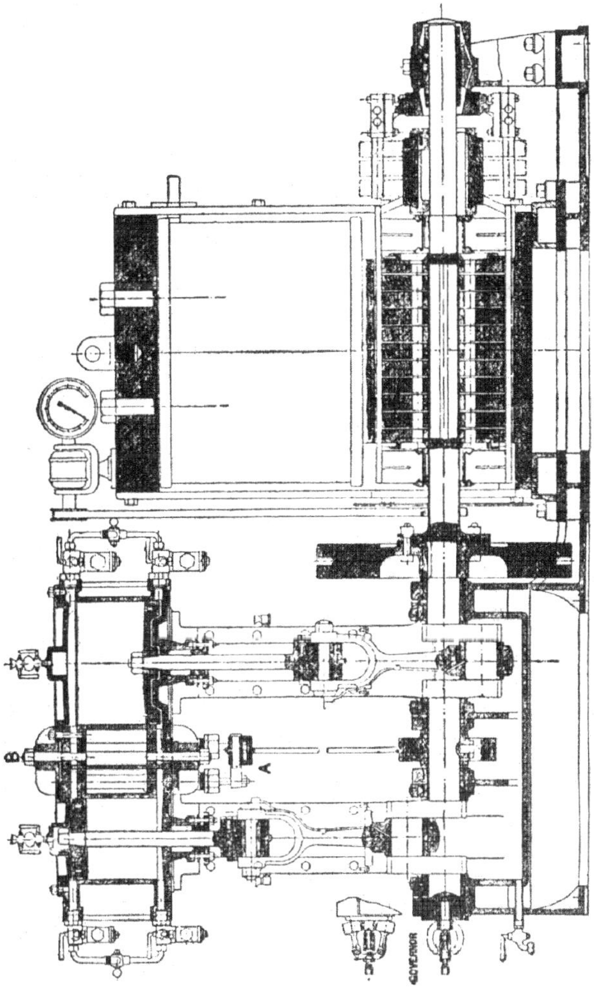

FIG. 167. ENGINES AND DYNAMO FOR MARINE SERVICE.

machines at the same time. The most recently fitted dynamos for the marine service are of the iron-clad type, the field coils and the armature being almost entirely surrounded by iron, to reduce to a minimum the leakage of magnetic lines of force which may affect compasses or chronometers in the neighborhood.

Ques. 616.—How are these dynamos usually driven?

Ans.—By vertical two-cylinder engines, generally compounded, although in some ships, where the steam-pressure is low, the engines are simple. All parts are carefully balanced and a heavy fly-wheel is fitted on the engine-shaft, at the dynamo end, which conduces to steady running. The speed is regulated by an isochronal governor fitted on the shaft.

Ques. 617.—Describe the construction of the armature.

Ans.—The armature-core is built up of thin disks of soft iron slipped over metal sleeves, which are keyed on the shaft. The disks are insulated from each other by thin sheets of asbestos paper, to prevent loss of energy and heating due to eddy currents, and are kept in place by clamping-plates and end-nuts. The conductors on the armature, which carry the current, are made up of copper wires, twisted together, and pressed to a rectangular section. They are insulated by a covering of varnished tape. Usually two lengths of bars are used. They are placed around the periphery of the armature, longitudinally, long and short bars alternating, their ends overhanging the core. All the ends at one end of the armature project the same distance. Projections are fitted into the core at intervals, which drive the conductor-bars. These projections are insulated by mica slips. The bars are kept in place by bands of steel or bronze binding wire, tightly wound on and soldered. Mica strips are placed under the bands to prevent injury to the insulation of the bars. Each bar is connected at each end by bent copper strips to another bar almost diametrically opposite to it, so that the whole of the bars and end-connections form one closed circuit. The projecting end of each long bar is also connected to the nearest commutator segment, the number of segments being equal to the number of long

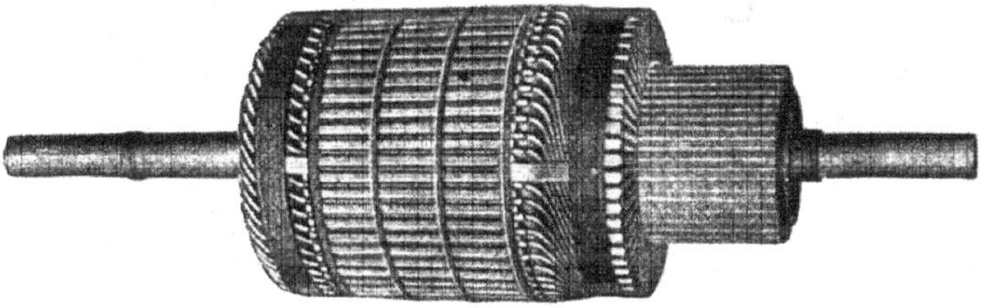

FIG. 168. ARMATURE

bars. Two or more pairs of brushes bear on the commutator, to collect the current, so that any brush may be lifted off without interrupting the circuit.

Ques. 618.—Describe the construction of the field-magnet coils.

Ans.—The field-magnet winding consists of shunt and series coils wound on a frame which fits over the upper pole-piece. The shunt coils are of small wire and high resistance. The ends of the wire are connected to the machine terminals. The greater part of the magnetism is due to these coils, so that at full speed, and when no current is being taken from the machine, the electric pressure is normal, that is, 80 or 100 volts. The series coils are formed of thick copper bars and convey the whole current generated. They provide additional magnetism, proportional to the current flowing in them, and so compensate for the additional

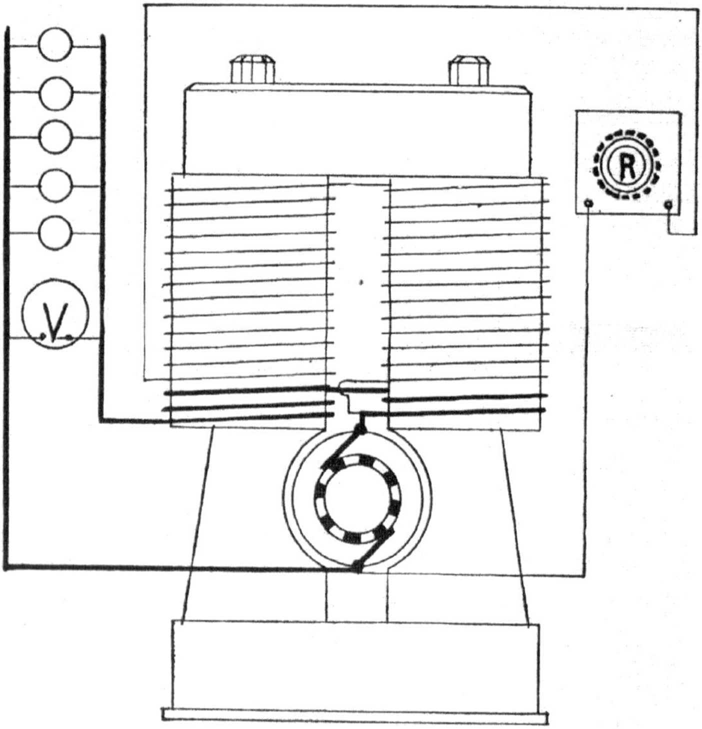

FIG. 169. COMPOUND WOUND DYNAMO.

pressure required to force this current through the machine. By the combination of the two sets of coils, the pressure is thus independent of the current, so long as the speed is constant. In the largest machines there are two distinct armature windings laid on side by side, the bars of the two windings alternating, as also do their respective commutator segments. The two windings are connected in parallel by the brushes, which all have a bearing rather wider than the angular width of two commutator segments.

Ques. 619.—In order to obtain satisfactory working, what should be done with the commutator occasionally?

Ans.—It should be turned up, by using a lathe slide-rest clamped to the bed-plate and running the engines as slowly as possible, and after turning, the commutator should be polished. This truing up is necessary in order to remove any flat places which are liable to form on the segments. The brushes also should be carefully filed to fit the commutator curve. The brushes must be carefully set in the holders, with all the tips of each set in a line, and the tips of the two sets bearing simultaneously on diametrically opposite commutator segments. Generally two segments are marked at their ends, with crosses, to assist in this adjustment.

Ques. 620.—How is the electric current carried to the different parts of the ship?

Ans.—By wires of the best copper, thoroughly insulated and protected from injury by being placed in wooden mouldings, or what is still better, iron tubes lined with insulating material. The junction boxes have safety fuses and connections, arranged in incombustible porcelain or lava blocks.

Ques. 621.—How are the lamps and motors arranged?

Ans.—The lamps are attached to substantial supports with good protection to the insulation of the wires at their connection. For exposed places extra globes or wire screens are provided to prevent breaking of the bulbs. The motors are fitted on substantial foundations, with switches for handling in convenient positions. The use of electric motors is becoming more and more general on board vessels as their convenience and freedom from waste is known. They can be used for working ventilating fans, etc., in confined spaces where the heat of steam would be objectionable. They also avoid the waste due to condensation, radiation and leakage in pipes, require very little attention when running and are always ready for starting.

Ques. 622.—What facilities are provided for pumping the water out of steam-ships in case of a serious leak?

Ans.—All steam-ships, including war-vessels, were formerly fitted with bilge-pumps worked direct from the main engines, and this is still the common practice in the mercantile marine. In addition to these pumps, the circulating pumps are fitted with bilge as well as sea connections, and in some of the larger vessels there are four centrifugal pumps which can be used for pumping out the bilges, each of these pumps having a capacity of at least 1,200 tons of water per hour.

Ques. 623.—What are some of the requirements of a reliable bilge-pumping outfit?

Ans.—The pump itself should be close to the bilge, but the engine for working it should if possible be at a high level, so as to be out of the reach of the water in case of its rising rapidly. Another point that should be kept in view is the provision of large engine-power for working the pumps. The valves for changing the suction of the centrifugal pumps from the sea to the bilge are, or at least should be, arranged to be worked from the starting platform, and to enable this to be done quickly in case of need, the valves in the sea and bilge-suction pipes

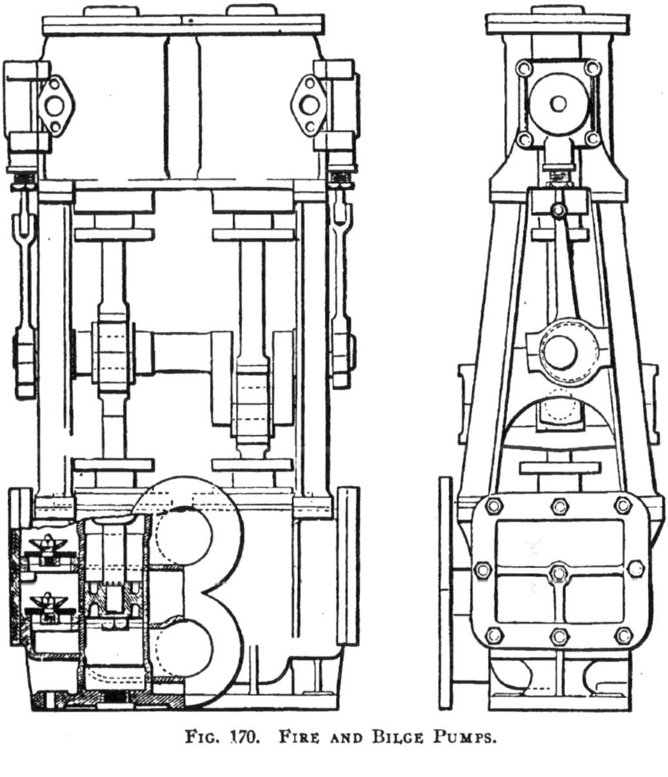

FIG. 170. FIRE AND BILGE PUMPS.

are often coupled together so that they may be worked by a single lever.

Ques. 624.—Describe the type of fire and bilge-pumping engines that are used to a large extent in the English navy.

Ans.—Each pumping engine consists of two double-acting pumps and two steam-cylinders, fitted with slide-valves, having very little lap, to insure the engines starting readily from any position of the cranks, economy in the use of steam being in these cases a minor consideration. In the large battle-ships and cruisers there are four of these pumps, two in each engine-room, each one of the four having a capacity of 80 to 120 tons of water per hour. The pumps are large enough to remove these quantities of water at a speed not exceeding 60 revolutions per minute, with a steam-pressure of two-thirds the maximum boiler-pressure, and they form a means of pumping water out of the ship, auxiliary to the main circulating pumps. They can be used for either fire service or for clearing the bilges of water.

Ques. 625.—Describe Friedmann's bilge-ejector.

Ans.—This apparatus is a modification of Giffard's injector, the number of nozzles being increased so as to give the steam several suction orifices instead of one. The steam is conducted to a tuyere about one-half the diameter of the steam-pipe, and then passes successively through a series of intermediate tuyeres, through which the water is drawn from the hold and expelled from the ship through the discharge. The device occupies little space and has considerable capacity, but its consumption of steam is large.

Ques. 626.—Describe the suction and discharge arrangements of fire and bilge pumps.

Ans.—They are fitted with separate suction-pipes leading to the following parts of the vessel: Forward and after ends of engine-room, with a continuation to the screw tunnel from the latter, main engine save-all, each boiler compartment, the main suction-pipe, salvage system of the vessel and to the sea. The valve-boxes and pipes are so arranged that each pump can draw from any of these parts. The pumps deliver water either overboard direct, to the engine-room or to the fire-main, a large air-vessel being fitted in connection with the latter.

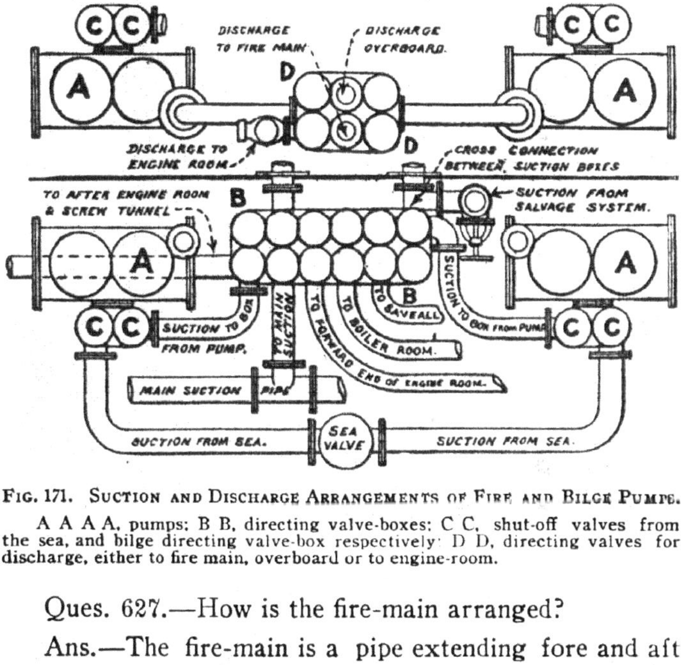

FIG. 171. SUCTION AND DISCHARGE ARRANGEMENTS OF FIRE AND BILGE PUMPS.
A A A A, pumps; B B, directing valve-boxes; C C, shut-off valves from the sea, and bilge directing valve-box respectively; D D, directing valves for discharge, either to fire main, overboard or to engine-room.

Ques. 627.—How is the fire-main arranged?

Ans.—The fire-main is a pipe extending fore and aft in the ship, with branches leading to different parts as required. Delivery-valves, with screwed nozzles for hose-connections, are located at various points in the fire-main. Non-return valves are fitted at the junction of delivery-pipes from the pumping engines.

CHAPTER IX

THE INDICATOR—PRINCIPLES OF THE INDICATOR

Ques. 628.—By whom was the indicator invented and first applied to the steam-engine?

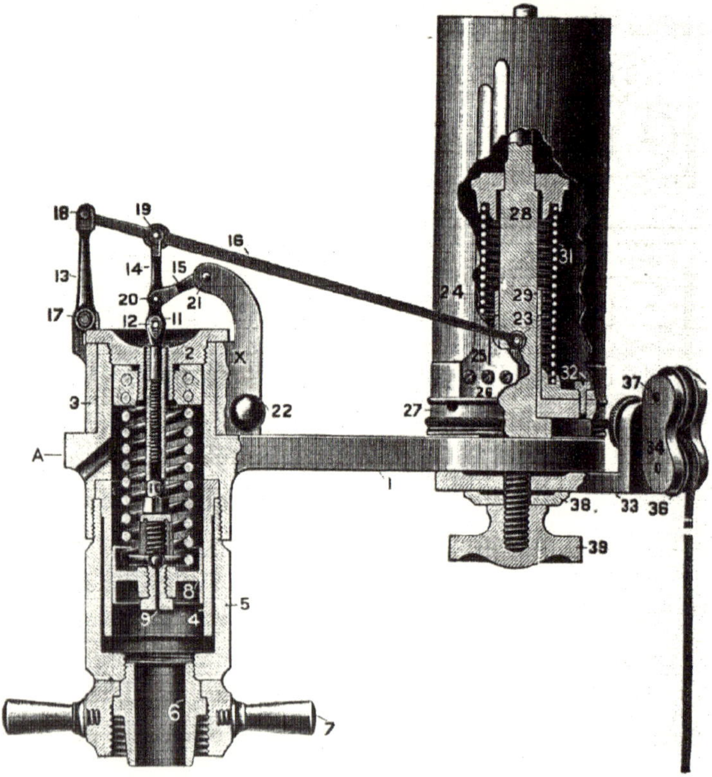

FIG. 172. SECTIONAL VIEW CROSBY INDICATOR.

Ans.—The indicator was invented and first applied to the steam-engine by James Watt, whose restless genius was not satisfied with a mere outside view of his engine as it was running, but he desired to know more about the action of the steam in the cylinder, its pressure at different portions of the stroke, the laws governing its expansion after being cut off, etc. Watt's indicator, although crude in its design and construction, contained embodied within it all of the principles of the modern instrument.

Ques. 629.—What are the principles governing the action of the indicator?

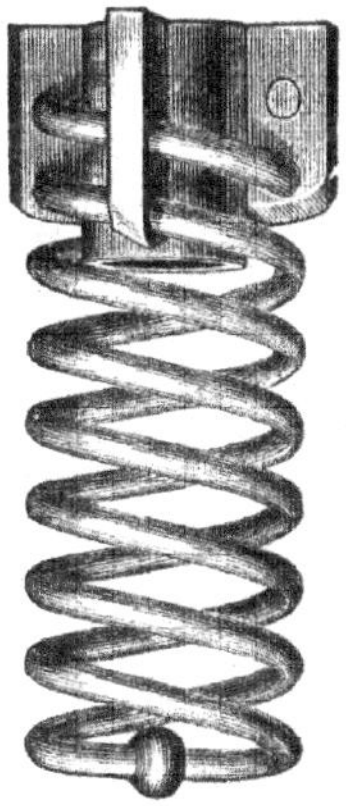

Fig. 173.
Crosby Indicator Spring.

Ans.—First, the pressure of the steam in the engine-cylinder throughout an entire revolution, against a small piston in the cylinder of the indicator, which in turn is controlled or resisted in its movement by a spring of known tension, so as to confine the stroke of the indicator piston within a certain small limit. Second, the stroke of the indicator piston is communicated by a multiplying mechanism of levers and parallel motion to a pencil moving in a vertical straight line, the distance through which the pencil moves being governed by the pressure in the engine-cylinder and the tension of the spring. Third, by the intervention of a reducing mechanism and a strong cord, the motion of the piston of the engine throughout an entire revolution is communicated to a small drum attached to and forming a part of the indicator. The movement of the drum is rotative and in a direction at right angles to the movement of the pencil. The forward stroke of the engine-piston causes the drum to rotate through part of a revolution and at the same time a clock-spring connected within the drum is wound up. On the return stroke the motion of the drum is reversed, and the tension of the spring returns the drum to its original position and also keeps the cord taut.

Ques. 630.—Describe in general terms the construction of an indicator.

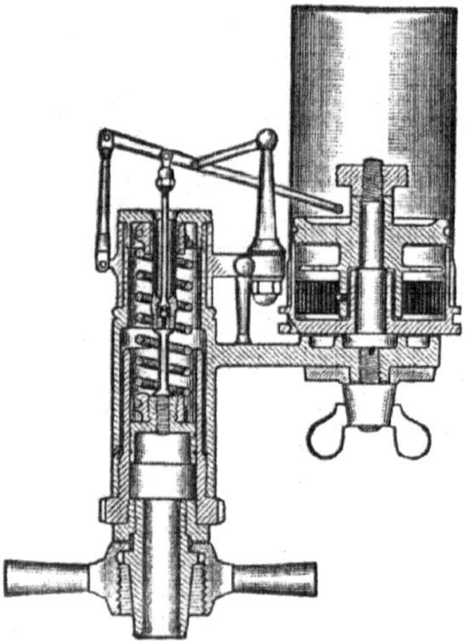

Fig. 174.
Sectional View Thompson Indicator.

Ans.—An indicator consists of a small cylinder, open to the atmosphere at the top and having its bottom end connected by suitable pipes and stop-cocks to both ends of the engine-cylinder in such a manner that the steam-pressure in either end may be caused to act upon the indicator piston, as required. The cylinder of the indicator stands vertical, and is of a known area, usually about one square inch. It contains a piston, upon which the steam acts only on the under side, the top of the cylinder being open to the atmosphere. The length of stroke of this piston is regulated and controlled by a steel spiral spring of known tension, which acts in resistance to the pressure of the steam. When the cock connecting the cylinders of the engine and indicator is closed, both ends of the indicator cylinder are open to atmospheric pressure, and the pencil, which is connected to the piston by a system of levers, stands at its neutral position.

Ques. 631.—Describe the construction and action of the spiral spring in connection with the indicator piston.

Ans.—These springs are made of different tensions in order to be suitable to different steam-pressures and speeds, and are numbered 20, 40, 60, etc., the number meaning that a pressure per square inch in the engine-cylinder corresponding to the number on the spring will cause a vertical movement of the pencil through a distance of one inch. Thus, if a No. 20 spring is used and the pressure in the cylinder at the commencement of the stroke is 20 pounds per square inch, the pencil will be raised one inch, or if the pressure is 30 pounds, the pencil will travel 1½ inch, and if there is a vacuum of 20 inches in the condenser, the pencil will drop ½ inch below the atmospheric line for the reason that 20 inches of vacuum correspond to a pressure of about 10 pounds less than atmospheric pressure or an absolute pressure of about 4 pounds. If a 60 spring is used a pressure of 60 pounds in the engine-cylinder will be required to raise the pencil one inch, or 90 pounds to raise it 1½ inch.

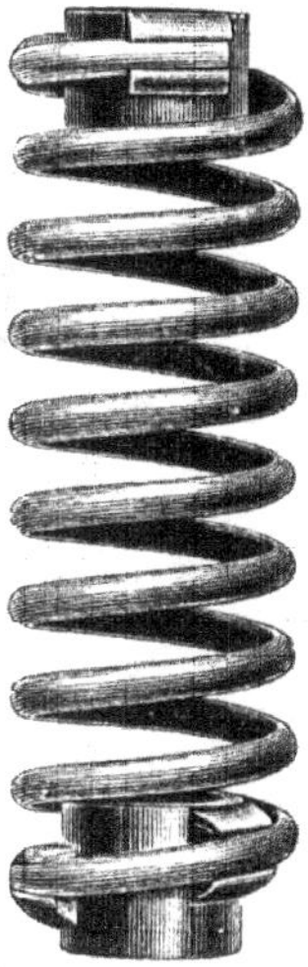

FIG. 175. THOMPSON INDICATOR SPRING.

Ques. 632.—Are these springs placed inside the cylinder in all types of indicators?

Ans.—The Ashcroft Manufacturing Company of New

York, makers of the well-known Tabor indicator, have recently introduced a new feature in indicator work by connecting the spring on top of the cylinder and in plain

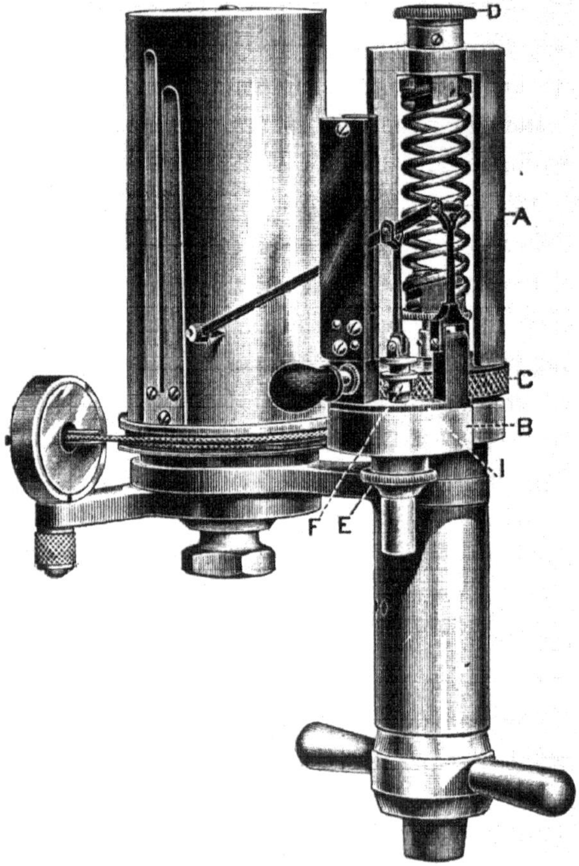

FIG. 176. IMPROVED TABOR INDICATOR WITH OUTSIDE CONNECTED SPRING. ASHCROFT MFG. CO., N. Y.

view of the operator. This arrangement removes the spring from the influence of direct contact with the steam, and it is subject only to the temperature of the surrounding atmosphere. It is claimed that as a result of this the accuracy of the spring is insured and that no allowance need to be made in its manufacture for expansion caused by the high temperature to which it is subject when located within the cylinder. Another good feature of this design is, that the spring can be easily removed without disconnecting any one part of the instrument in case it is desired to change springs.

Ques. 633.—What precautions should be observed in attaching the indicator to an engine-cylinder?

Ans.—The main requirements in these connections are that the holes shall not be drilled near the bottom of the cylinder where water is likely to find its way into the pipes, neither should they be in a location where the inrush of steam from the ports will strike them directly, nor where the edge of the piston is liable to partly cover them when at its extreme travel. An engineer before he undertakes to indicate an engine should satisfy himself that all these requirements are fulfilled. Otherwise he is not likely to obtain a true diagram. The cock supplied with the indicator is threaded for one-half inch pipe, and unless the engine has a very long stroke it is the practice to bring the two end connections together at the side or top of the cylinder and at or near the middle of its length, where they can be connected to a three-way cock. The pipe connections should be as short and as free from elbows as possible, in order that the steam may strike the indicator piston as nearly as possible at the same moment that it acts upon the engine-piston. These pipes should always be thoroughly blown out and cleaned, by allowing the steam to blow through the open three-way cock during several revolutions of the engine before connecting the indicator. If this is not done there is a moral certainty that dirt and grit will get into the cylinder of the indicator and cause it to work badly and give diagrams that are misleading.

Ques. 634.—How is an indicator diagram or card drawn?

FIG. 177. THREE-WAY COCK.

Ans.—To the outside of the drum a piece of blank paper of suitable size is attached and held in place by two clips. Upon this paper the pencil in its motion up and down traces a complete diagram of the pressures and other interesting events transpiring within the engine-cylinder during the revolution of the engine. In fact, the diagram traced upon the paper is the compound result of two concurrent movements. First, that of the pencil caused by the pressure of the steam against the indicator piston; second, that of the paper drum caused by, and coincident with the motion of the engine-piston.

Ques. 635.—How is the atmospheric line drawn?

Ans.—By holding the pencil to the paper, and causing the drum to be rotated, when the pencil stands at its neutral position, that is with the steam shut off from the indicator cylinder.

Ques. 636.—What is meant by the term atmospheric line?

Ans.—The atmospheric line is a horizontal line drawn on the diagram and means the line of atmospheric pressure. If the engine is a non-condensing engine the pencil in tracing the diagram will, or at least should not fall below the atmospheric line at any point, but will on the return stroke trace a line called the line of back pressure at a distance more or less above the atmospheric line and very nearly parallel with it. If the engine is a condensing engine the pencil will drop below the atmospheric line while tracing the line of back pressure on the diagram, and the distance this line is below the atmospheric line will depend upon the number of inches of vacuum in the condenser.

Ques. 637.—Is the atmospheric line a necessary part of an indicator diagram?

Ans.—The atmospheric line is a very important factor in the study of the diagram.

Ques. 638.—How are the dimensions of the diagram regulated?

Ans.—It is a convenient practice to select a spring numbered one-half of the boiler-pressure as, for instance, suppose gauge-pressure or boiler-pressure is 200 pounds per square inch, then a 100 spring would give a diagram 2 inches in height, which is a convenient height. As to the length of the diagram, this is regulated by adjustment

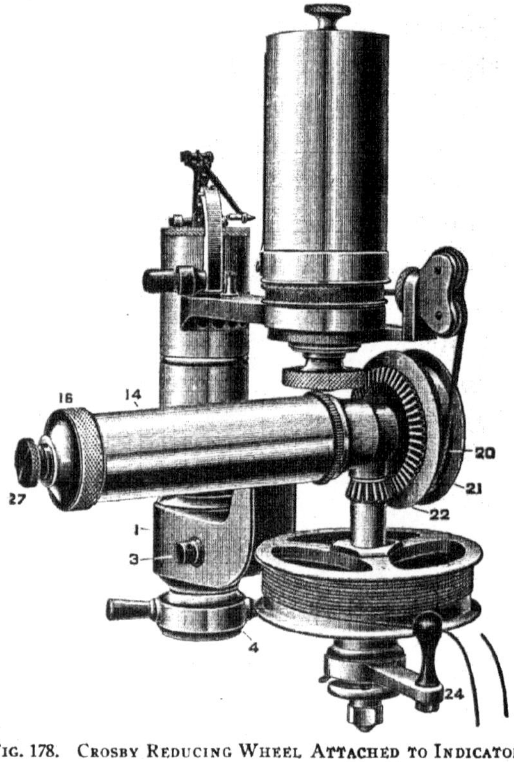

FIG. 178. CROSBY REDUCING WHEEL ATTACHED TO INDICATOR.

of the cord in its travel, by means of the reducing wheel. Any length of diagram up to four inches may be obtained, but two and a half to three inches is a very good length for analysis.

Ques. 639.—How is the motion of the crosshead of

the engine reduced and utilized for rotating the drum of the indicator?

Ans.—There are various mechanisms used for this purpose. Probably the only practically universal

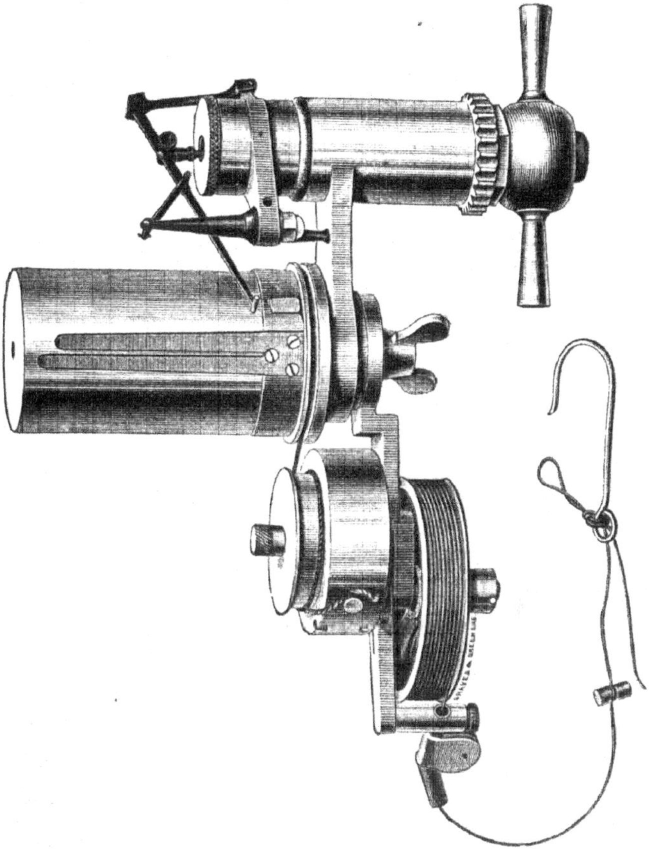

FIG. 179. THOMPSON INDICATOR WITH REDUCING WHEEL ATTACHED.

mechanism for reducing the motion of the crosshead is the reducing wheel, a device in which, by the employment of gears and pulleys of different diameters, the motion is reduced to within the compass of the drum, and the device is applicable to almost any make of engine, whether of high or low speed Some makers of indicators attach the reducing wheel directly to the indicator, thus producing a neat and very convenient arrangement.

Ques. 640.--Describe the construction of the wooden pendulum for reducing the motion.

Ans.—It consists of a flat strip of pine or other light

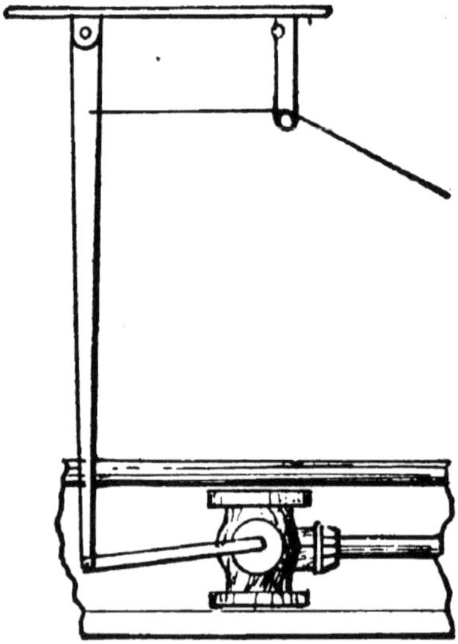

WOODEN PENDULUM, REDUCING MOTION

wood of a length not less than one and a half times the stroke of the engine, and if made longer it will be better. It should be from ¾ to ⅞ inch thick and have an average width of about 4 inches. If the engine to be indicated is horizontal the bar or pendulum is to be pivoted at a fixed point directly above and in line with the side of the cross-head, as that is generally the most convenient point of attachment. The pivot can be fixed to a permanent standard bolted to the frame of the engine or it may be secured to the ceiling of the room or even to a post fastened to the floor. If the engine is vertical the bar can be pivoted to the wall of the room or a strong post firmly secured to the floor. The connection with the crosshead is best accomplished by means of a short bar or link. A convenient length for this bar is one-half the stroke of the engine.

Ques. 641.—When the short bar is one-half the length of the stroke, how is the correct point for the location of the pivot for the pendulum found?

Ans.—Place the engine on the center with the crosshead at the end of the stroke towards the crank. Then having previously bored a hole for the pivot in one end of the pendulum bar and in the other end a hole for connecting with the link, suspend the pendulum by a temporary pin, as a large wood screw, directly above and in line with the stud or bolt hole which has previously been tapped into the crosshead at any convenient point. The pendulum should be temporarily suspended at such a height that when it hangs perpendicular the hole in its lower end will line up accurately with the hole or stud in the crosshead. Now swing the pendulum in either direction a distance equal to the length of the link (one-half the stroke of the engine) from the crosshead connection and note the distance that the bottom hole is above a straight edge laid horizontal and in line with the center of the stud in the crosshead. This will give the total vibration of the free end of the link from a line parallel with the line of the engine and the permanent location of the pivot should be one-half of this distance below the temporary point of suspension. This will allow the link to vibrate equally above and below the center of its connection with the crosshead.

Ques. 642.—How is the correct point of attachment of the cord to the pendulum found?

Ans.—The cord can be attached to the pendulum at a point near the pivot, which will give the desired length of diagram. This point can be determined by multiplying the length of the pendulum by the desired length of diagram and dividing the product by the stroke. For convenience these terms should be expressed in inches. Thus, assume stroke of engine to be 48 inches, length of pendulum 1½ times length of stroke = 72 inches. Desired length of diagram 3 inches. Then $72 \times 3 \div 48 = 4.5$ inches, which is the distance from center of pivot to point of connection for the cord. This can be either a small hole bored through the pendulum or a wood screw to which the cord can be attached. From this point the cord should be led over a guide pulley located at such height that when the pendulum is vertical the cord will leave it at right angles. After leaving the guide pulley the cord can be carried at any angle desired.

Ques. 643.—How should the indicator be cared for?

Ans.—The indicator should be cleaned and oiled both before and after using. The best material for wiping it is a clean piece of old soft muslin of fine texture, as there is not so much liability of lint sticking to or getting into the small joints. Use good clock oil for the joints and springs, and before taking diagrams it is a good practice to rub a small portion of cylinder oil on the piston and the inside of the cylinder, but when about to put the instrument away these should be oiled with clock oil also.

Ques. 644.—How may the cord be adjusted to proper length?

Ans.—None but the best cord should be used for connecting the paper drum with the reducing motion, as a cord that is liable to stretch will cause trouble. After the indicator has been screwed on to the cock connecting with the pipe, the cord must be adjusted to the proper length before hooking it on to the drum. This must be done while the engine is running, by taking hold of the loop on the cord connected with the reducing motion with one hand, and with the other hand grasp the hook on the short cord attached to the drum, then by holding the two ends near each other during a revolution or two it will be seen whether the long cord needs to be shortened or lengthened.

Ques. 645.—What precautions are necessary in regard to the paper and pencil in order to secure a truthful diagram?

Ans.—Care should be exercised in placing the paper on the drum to see that it is stretched tight and firmly held by the clips. The pencil point having been first sharpened by rubbing it on a piece of fine emery cloth or sand paper should be adjusted by means of the pencil stop with which all indicators should be provided, so that it will have just sufficient bearing against the paper to make a fine, plain mark. If the pencil bears too hard on the paper it will cause unnecessary friction and the diagram will be distorted. The best method of ascertaining this fact and also whether the travel of the drum is equally divided between the stops, is to place a blank diagram on the drum, connect the cord and while the engine makes a revolution hold the pencil against the paper. Then unhook the cord, remove the paper and if the travel of the drum is not divided correctly it can be changed.

Ques. 646.—Describe the process of taking an indicator diagram.

Ans.—Place a fresh blank on the drum, being careful to keep the pencil out of contact with it, connect the cord, open the cock admitting steam to the indicator and after the pencil has made a few strokes to allow the cylinder to become warmed up, then gently swing it around to the paper drum and hold in there while the engine makes a complete revolution. Then move the pencil clear of the paper, close the cock and unhook the cord. Now trace the atmospheric line by holding the pencil against the paper while the drum is revolved by hand. This method of tracing the atmospheric line is preferable to that of tracing it immediately after closing the cock and while the drum is still being moved by the engine, for the reason that there is not so much liability of getting the atmospheric line too high owing to the presence of a slight pressure of steam remaining under the indicator piston for a second or two just after closing the cock; also the line drawn by hand will be longer than one drawn while the drum is moved by the motion of the engine and will therefore be more readily distinguished from the line of back pressure.

Ques. 647.—What other details should be observed in the taking of indicator diagrams?

Ans.—As soon as the diagrams are taken the following data should be noted upon them: The end of the cylinder, whether head or crank; boiler-pressure, and time when taken. Other data can be added afterwards.

Ques. 648.—What needed changes in the cut-off of a Corliss engine, as shown by a diagram, may be made while the engine is running?

Ans.—If the engine is an automatic cut-off of the Corliss type and the point of cut-off on one end does not coincide with the other, the difference can generally be adjusted while the engine is running by changing the length of the rods extending from the governor to the tripping device. These rods are, or should be fitted with right and left threads on the ends for this purpose. Any changes in the valves, such as giving them more lead, compression, etc., and which necessitates changing the length of the reach rods connecting them with the wrist plate, will have to be made while the engine is stopped, although with slow-speed engines and the exercise of caution it is possible to make alterations in these rods while the engine is running.

Ques. 649.—What important details will a truthful indicator diagram show?

Ans.—First, the pressure of the steam against the piston of the engine at any point in the stroke during a complete revolution; second, diagrams from a condensing engine show the amount of vacuum that is being maintained in the condenser, measured from the line of perfect vacuum; third, the point of cut-off is clearly shown, also the point in the return stroke at which compression begins; fourth, the expansion curve, and how near it approaches the theoretical expansion curve; fifth, any fault in the setting of the valves is clearly shown on the diagram; sixth, diagrams taken from the different cylinders of a compound or stage expansion engine may be combined in such a manner as to show whether or not the cylinders are properly proportioned, and whether the steam is being distributed correctly.

Ques. 650.—What is absolute pressure?

Ans.—Pressure reckoned from a perfect vacuum. It equals the boiler-pressure plus the atmospheric pressure.

Ques. 651 —What is boiler-pressure or gauge-pressure?

Ans.—Pressure above the atmospheric pressure as shown by the steam gauge.

Ques. 652.—What is initial pressure?

Ans.—Pressure in the cylinder at the beginning of the stroke.

Ques. 653.—What is meant by terminal pressure (T. P.)?

Ans.—The pressure that would exist in the cylinder at the end of the stroke provided the exhaust valve did not open until the stroke was entirely completed. It may be graphically illustrated on the diagram by extending the expansion curve by hand to the end of the stroke. It is found theoretically by dividing the pressure at point of cut-off by the ratio of expansion. Thus, absolute pressure at cut-off=100 pounds, ratio of expansion=5; then 100÷5=20 pounds, absolute terminal pressure.

Ques. 654.—What is mean effective pressure (M. E. P.)?

Ans.—The average pressure acting upon the piston throughout the stroke minus the back pressure.

Ques. 655.—What is back pressure?

Ans.—Pressure which tends to retard the forward stroke of the piston. Indicated on the diagram from a non-condensing engine by the height of the back pressure line above the atmospheric line. In a condensing engine the degree of back pressure is shown by the height of the back pressure line above an imaginary line representing the pressure in the condenser corresponding to the degree of vacuum in inches, as shown by the vacuum gauge.

Ques. 656.—What is total or absolute back pressure?

Ans.—Total or absolute back pressure, in either a condensing or non-condensing engine, is that indicated on the diagram by the height of the line of back pressure above the line of perfect vacuum.

Ques. 657.—How is the line of perfect vacuum drawn on an indicator diagram?

Ans.—The line of perfect vacuum is drawn parallel with the atmospheric line and at a distance below the latter, representing 14.7 pounds, as measured by the scale corresponding to the spring that was used in taking the diagram. Different scales are supplied for the different springs used.

Ques. 658.—What is meant by ratio of expansion?

Ans.—The proportion that the volume of steam in the

cylinder at point of release bears to the volume at cut-off. Thus, if the point of cut-off is at one-fifth of the stroke, and release does not take place until the end of the stroke, the ratio of expansion, or in other words, the number of expansions, is 5. When the T. P. is known the ratio of expansion may be found by dividing the initial pressure by the T. P.

Ques. 659.—What is meant by wire drawing?

Ans.—When through insufficiency of valve opening, contracted ports or throttling governor, the steam is prevented from following up the piston at full initial pressure until the point of cut-off is reached, it is said to be wire drawn. It is indicated on the diagram by a gradual inclination downwards of the steam line from the admission line to the point of cut-off. Too small a steam pipe from boiler to engine will also cause wire drawing and fall of pressure.

Ques. 660.—What is condenser pressure?

Ans.—Condenser pressure may be defined as the pressure existing in the condenser of an engine, caused by the lack of a perfect vacuum. As, for instance, with a vacuum of 25 inches there will still remain the pressure due to the 5 inches which is lacking. This will be about 2.5 pounds.

Ques. 661.—What is absolute zero?

Ans.—Absolute zero has been fixed by calculation at 461.2 degrees below the zero of the Fahrenheit scale.

Ques. 662.—What is piston displacement?

Ans.—The space or volume swept through by the

piston in a single stroke. Found by multiplying the area of piston by length of stroke.

Ques. 663.—What is piston clearance?

Ans.—The distance between the piston and cylinder head when the piston is at the end of the stroke.

Ques. 664.—What is steam clearance, ordinarily termed clearance?

Ans.—The space between the piston at the end of the stroke and the valve face. It is reckoned in per cent of the total piston displacement.

Ques. 665.—What is the meaning of the expression horse-power as applied to a steam-engine?

Ans.—33,000 pounds raised one foot high in one minute of time.

Ques. 666.—What is indicated horse-power (I. H. P.)?

Ans.—The horse-power as shown by the indicator diagram. It is found as follows: Area of piston in square inches × M. E. P. × piston speed in feet ÷ 33,000.

Ques. 667.—What is meant by the term piston speed?

Ans.—The distance in feet traveled by the piston in one minute. It is the product of twice the length of stroke expressed in feet multiplied by the number of revolutions per minute.

Ques. 668.—What is net horse-power?

Ans.—I. H. P. minus the friction of the engine.

Ques. 669.—What is compression?

Ans.—The action of the piston as it nears the end of the stroke, in reducing the volume and raising the pressure of the steam retained in the cylinder ahead of the piston by the closing of the exhaust valve.

Ques. 670.—What is Boyle's law of expanding gases?

Ans.—"The pressure of a gas at a constant temperature varies inversely as the space it occupies." Thus, if a given volume of gas is confined at a pressure of 50 pounds per square inch and it is allowed to expand to twice its volume, the pressure will fall to 25 pounds per square inch.

Ques. 671.—What is an adiabatic curve?

Ans.—A curve representing the expansion of a gas which loses no heat while expanding. Sometimes called the curve of no transmission.

Ques. 672.—What is an isothermal curve?

Ans.—A curve representing the expansion of a gas having a constant temperature but partially influenced by moisture, causing a variation in pressure according to the degree of moisture or saturation. It is also called the theoretical expansion curve.

Ques. 673.—What is the expansion curve?

Ans.—The curve traced upon the diagram by the indicator pencil showing the actual expansion of the steam in the cylinder.

Ques. 674.—What is power?

Ans.—The rate of doing work, or the number of foot pounds exerted in a given time.

Ques. 675.—What is the unit of work?

Ans.—The foot pound, or the raising of one pound weight one foot high.

Ques. 676.—Define the first law of motion.

Ans.—All bodies continue either in a state of rest or of uniform motion in a straight line, except in so far as they may be compelled by impressed forces to change that state.

Ques. 677.—What is work?

Ans.—Mechanical force or pressure can not be considered as work unless it is exerted upon a body and causes that body to move through space. The product of the pressure multiplied by the distance passed through and the time thus occupied is work.

Ques. 678.—What is momentum?

Ans.—Force possessed by bodies in motion, or the product of mass and density.

Ques. 679.—What is the meaning of the word dynamics?

Ans.—The science of moving powers or of matter in motion, or of the motion of bodies that mutually act upon each other.

Ques. 680.—What is force?

Ans.—That which alters the motion of a body or puts in motion a body that was at rest.

Ques. 681.—What is the maximum theoretical duty of steam?

Ans.—The maximum theoretical duty of steam is the product of the mechanical equivalent of heat, viz., 778 foot pounds multiplied by the total heat units in a pound of steam. Thus, in one pound of steam at 212 degrees reckoned from 32 degrees the total heat equals 1,146.6 heat units. Then $778 \times 1,146.6 = 892,054.8$ foot pounds=maximum duty.

Ques. 682.—What is steam efficiency?

Ans.—Steam efficiency may be expressed as follows:

$\frac{\text{Heat converted into useful work}}{\text{Heat expended}}$ and maximum efficiency can only be attained by using steam at as high an initial pressure as is consistent with safety, and at as large a ratio of expansion as possible.

Ques. 683.—What is meant by the term efficiency of the plant as a whole?

Ans.—Efficiency of the plant as a whole includes boiler and engine efficiency, and is to be figured upon the basis of $\frac{\text{Heat converted into useful work}}{\text{Calorific or heat value of fuel}}$

Ques. 684.—What is the horse-power constant of an engine?

Ans.—The horse-power constant of an engine is found by multiplying the area of the piston in square inches by the speed of the piston in feet per minute and dividing the product by 33,000. It is the power the engine would develop with one pound mean effective pressure. To find the horse-power of the engine, multiply the M. E. P. of the diagram by this constant.

Ques. 685.—What is meant by the expression steam consumption per horse-power per hour?

Ans.—The weight in pounds of steam exhausted into the atmosphere or into the condenser in one hour divided by the horse-power developed. It is determined from the diagram by selecting a point in the expansion curve just previous to the opening of the exhaust-valve and measuring the absolute pressure at that point. Then the piston displacement up to the point selected, plus the clearance space, expressed in cubic feet, will give the volume of steam in the cylinder, which multiplied by the weight per cubic foot of steam at the pressure as measured will give the weight of steam consumed during one stroke. From this should be deducted the steam saved by compression as shown by the diagram, in order to get a true measure of the economy of the engine. Having thus determined the weight of steam consumed for one stroke, multiply it by twice the number of strokes per minute and by 60, which will give the total weight consumed per hour. This divided by the horse-power will give the rate per horse-power per hour.

Ques. 686.—What is cylinder condensation and reëvaporation?

Ans.—When the exhaust-valve opens to permit the exit of the steam there is a perceptible cooling of the walls of the cylinder, especially in condensing engines when a high vacuum is maintained. This results in more or less condensation of the live steam admitted by the opening of the steam-valve; but if the exhaust-valve is caused to close at the proper time so as to retain a portion of the steam to be compressed by the piston on the return stroke, a considerable portion of the water caused by condensation will be reëvaporated into steam by the heat and consequent rise in pressure caused by compression.

Ques. 687.—What are ordinates, as applied to indicator diagrams?

Ans.—Parallel lines drawn at equal distances apart across the face of the diagram and perpendicular to the atmospheric line. They serve as a guide to facilitate the measurement of the average forward pressure throughout the stroke, or the pressure at any point of the stroke if desired.

Ques. 688.—What is a throttling governor?

Ans.—A governor that is used to regulate the speed of engines having a fixed cut-off. The governor controls the position of a valve in the steam-pipe, opening or clos-

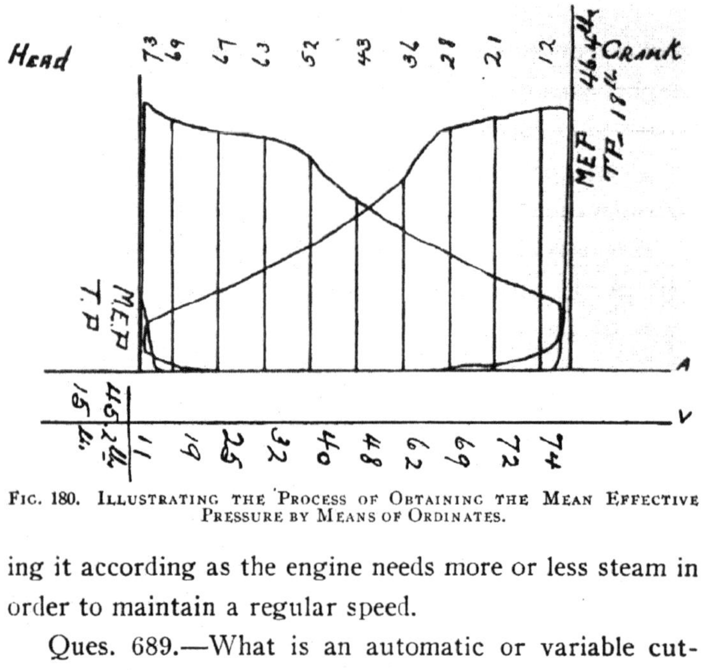

FIG. 180. ILLUSTRATING THE PROCESS OF OBTAINING THE MEAN EFFECTIVE PRESSURE BY MEANS OF ORDINATES.

ing it according as the engine needs more or less steam in order to maintain a regular speed.

Ques. 689.—What is an automatic or variable cut-off engine?

Ans.—In engines of this type the full boiler pressure is constantly in the valve chest and the speed of the engine is regulated by the governor controlling the point of cut-off, causing it to take place earlier or later according as the load on the engine is lighter or heavier.

Ques. 690.—What is a fixed cut-off?

Ans.—This term is applied to engines in which the point of cut-off remains the same regardless of the load, the speed being regulated by a throttling governor.

Ques. 691.—What is an adjustable cut-off?

Ans.—One in which the point of cut-off may be regulated or adjusted by hand by means of a hand wheel and screw attached to the valve stem, the supply of steam being regulated by a throttling governor.

Ques. 692.—What is an isochronal or shaft governor?

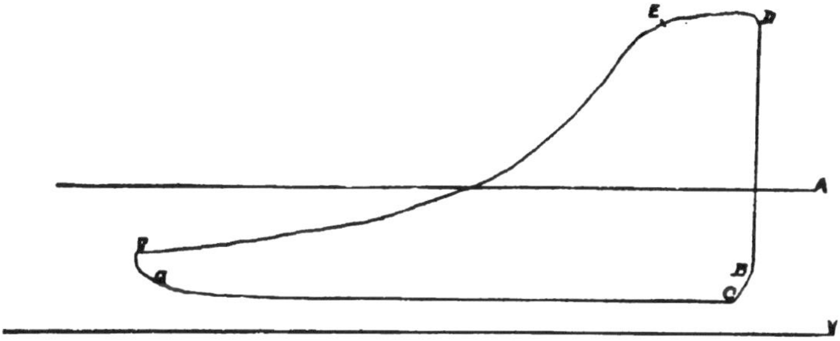

FIG. 181. INDICATOR DIAGRAM TAKEN FROM A CONDENSING ENGINE.

A, atmospheric line. V, line of perfect vacuum. B to D, admission line. D to E, steam line. E, point of cut-off. E to F, expansion line. F to G, exhaust. G to C, line of back pressure; and from C to B shows compression.

Ans.—This device in which the centrifugal and centripetal forces are utilized, as in the fly-ball governor, is generally applied to automatic cut-off engines having reciprocating or slide valves. It is attached to the crank shaft and its function is to change the position of the eccentric, which is free to move across the shaft within certain prescribed limits, but is at the same time attached to the governor. The angular advance of the eccentric is thus increased or diminished; in fact is entirely under the control of the governor, and cut-off occurs earlier or later according to the demands of the load on the engine.

Ques. 693.—If the valves of an engine are properly adjusted and the distribution of the steam is approximately correct, what particular features should characterize an indicator diagram taken from it?

Ans.—First, the admission line at the beginning of the stroke should be perpendicular to the atmospheric line; second, the steam line, as it is called, extending from the beginning of the stroke to the point of cut-off, should be

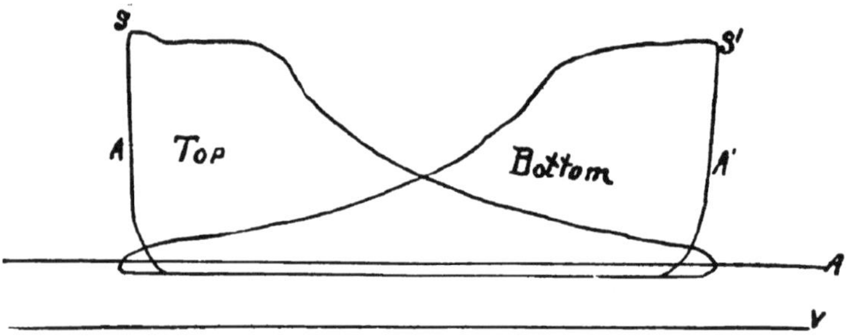

FIG. 182. DIAGRAM SHOWING INSUFFICIENT LEAD.

parallel with the atmospheric line; third, the point of cut-off should be sharply defined; fourth, the expansion curve, extending from the point of cut-off to the point of release, should conform as near as possible with the isothermal curve, which can easily be applied to any diagram; fifth, the exhaust line, extending from point of release to that point in the return stroke where compression begins, should be parallel with and practically coincident with the atmospheric line, if the engine is non-condensing, or if the engine be a condensing engine, this line should approach within a few pounds of the line of perfect vacuum.

Ques. 694.—If the admission line inclines inward from the perpendicular, what defect in the valve setting is indicated?

Ans.—Insufficient lead.

Ques. 695.—How is wire drawing of the steam detected by the indicator diagram?

Ans.—By the downward inclination of the steam line toward the point of cut-off.

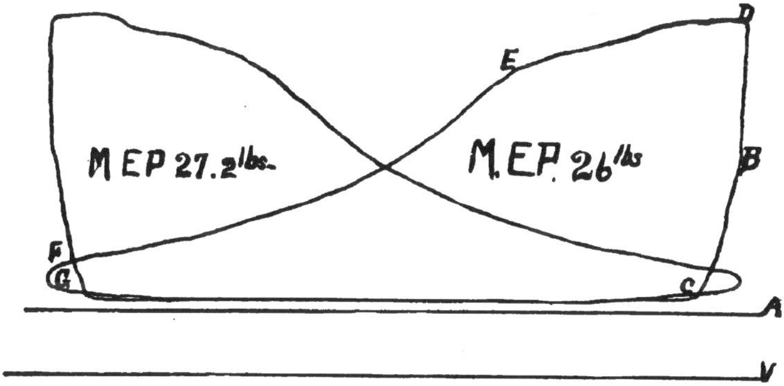

FIG. 183. DIAGRAM SHOWING EFFECTS OF WIRE DRAWING THE STEAM.

Ques. 696.—What is a very necessary factor in the calculation of the horse-power of an engine as shown by a diagram taken from it?

Ans.—The mean effective pressure.

Ques. 697.—How is the M. E. P. of a diagram ascertained?

Ans.—There are two methods commonly used. First, by means of ordinates, and secondly, by the use of the planimeter.

Ques. 698.—Describe the method of finding the M. E. P. by ordinates.

Ans.—The process consists in drawing any convenient number of vertical lines perpendicular to the atmospheric line across the face of the diagram, spacing them equally, with the exception of the two end spaces, which should be one-half the width of the others, for the reason that the ordinates stand for the centers of equal spaces. This is an important matter, and should be thoroughly under-

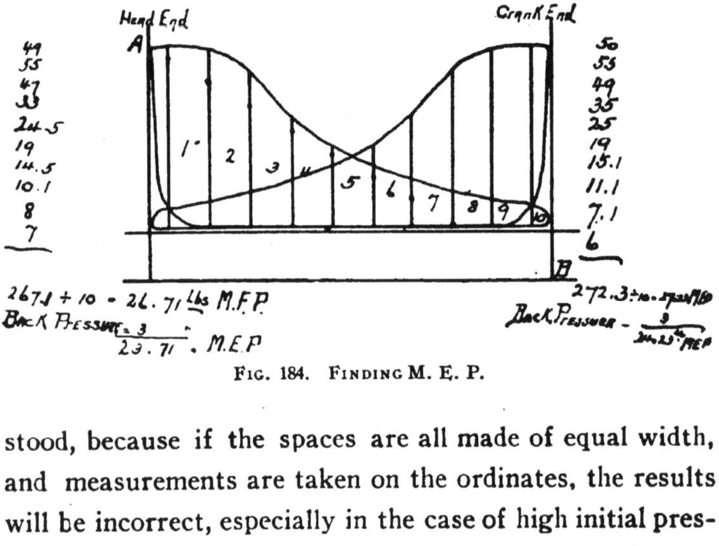

FIG. 184. FINDING M. E. P.

stood, because if the spaces are all made of equal width, and measurements are taken on the ordinates, the results will be incorrect, especially in the case of high initial pressure and early cut-off, following which the steam undergoes great changes. If the spaces are all made equal, the measurements will require to be taken in the middle of them, and errors are liable to occur, whereas if spaced as before described, the measurements can be made on the ordinates, which is much more convenient and will insure correct results. Any number of ordinates can be drawn, but ten is the most convenient and is amply sufficient, except in case the diagram is excessively long.

Ques. 699.—Having succeeded in drawing the ordinates across the face of the diagram, what is the next step?

Ans.—The pressure represented by each line is measured from the exhaust line to the steam line, and so on,

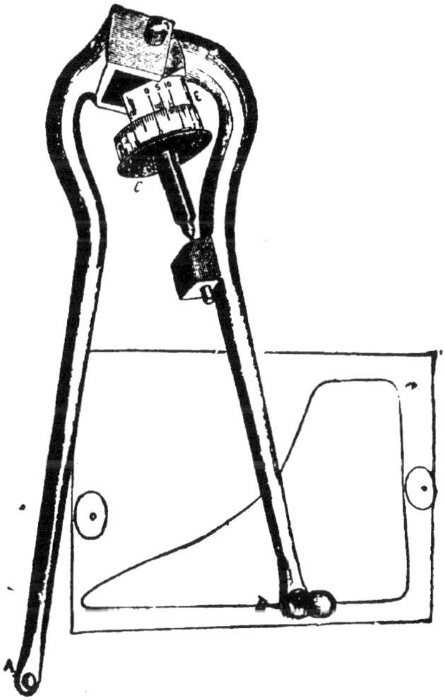

FIG. 185. PLANIMETER.

along the expansion curve throughout the length of the diagram, using for this purpose the scale adapted to the spring used, and having thus obtained measurements on each line, add all together and divide the sum total by the number of lines, which will give the mean forward pressure. To obtain the mean effective pressure, deduct the back pressure, which is represented by the distance of the exhaust line above the atmospheric line in a non-condensing engine, and in a condensing engine the back pressure is measured from the line of perfect vacuum.

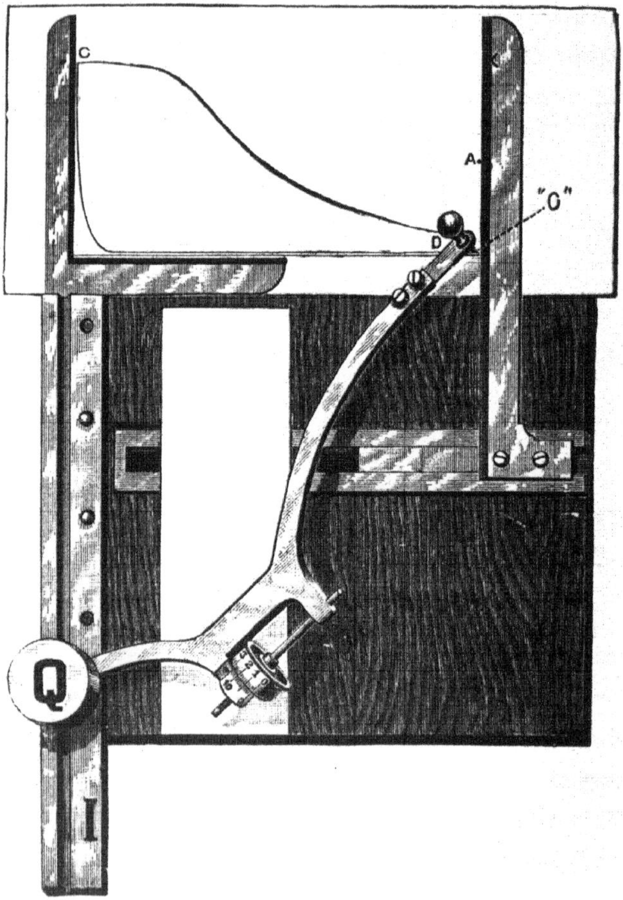

FIG. 186. COFFIN AVERAGER OR PLANIMETER.

Ques. 700.—What is a planimeter?

Ans.—The planimeter is an instrument which will

accurately measure the area of any plane surface, no matter how irregular the outline or boundary line is.

Ques. 701.—What is the main requirement in ascertaining the M. E. P. of a diagram?

Ans.—The prime requisite in making power calculations from indicator diagrams is to obtain the average height or width of the diagram, supposing it were reduced to a plain parallelogram instead of the irregular figure which it is.

Ques. 702.—What advantage is gained by using the planimeter in measuring diagrams?

Ans.—It shows at once the area of the diagram in square inches and decimal fractions of a square inch, and when the area is thus known it is an easy matter to obtain the average height by simply dividing the area in inches by the length of the diagram in inches. Having ascertained the average height of the diagram in inches or fractions of an inch the mean or average pressure is found by multiplying the height by the scale. Or the process may be made still more simple by first multiplying the area, as shown by the planimeter in square inches and decimals of an inch, by the scale and dividing the product by the length of the diagram in inches. The result will be the same as before, and troublesome fractions will be avoided.

Ques. 703.—Having obtained the M. E. P., as shown by the diagram, how may the horse-power developed by the engine be ascertained?

Ans.—The area of the piston (minus one-half the area of rod) multiplied by the M. E. P., as shown by the diagram, and this product multiplied by the number of feet traveled by the piston per minute (piston speed) will give the number of foot pounds of work done by the engine each minute, and if this product be divided by 33,000, the quotient will be the indicated horse-power (I. H. P.) developed by the engine.

Ques. 704.—Mention two important factors in calculations of steam consumption.

Ans.—In calculating the steam consumption of an engine, two very important factors must not be lost sight of, viz., clearance and compression. Especially is this the case in regard to clearance when there is little or no compression, for the reason that the steam required to fill the clearance space at each stroke of the engine is practically wasted, and all of it passes into the atmosphere or the condenser, as the case may be, without having done any useful work except to merely fill the space devoted to clearance. On the other hand, if the exhaust valve is closed before the piston completes the return stroke, the steam then remaining in the cylinder will be compressed into the clearance space and can be deducted from the total volume which, without compression, would have been exhausted at the terminal pressure.

Ques. 705.—When, owing to light load and early cut-off, the expansion curve drops below the line of back pressure, how must the area of the diagram be calculated?

Ans.—The area of the loop below the back pressure line must be subtracted from the remainder of the diagram. If the planimeter is used, the instrument will make the subtraction automatically, but if the diagram is divided into parts by ordinates, the pressure shown by the ordinates in the lower loop must be subtracted from that shown by the loop above the back pressure line in order to ascertain the M. E. P. or average pressure.

Ques. 706.—What is meant by the adiabatic curve?

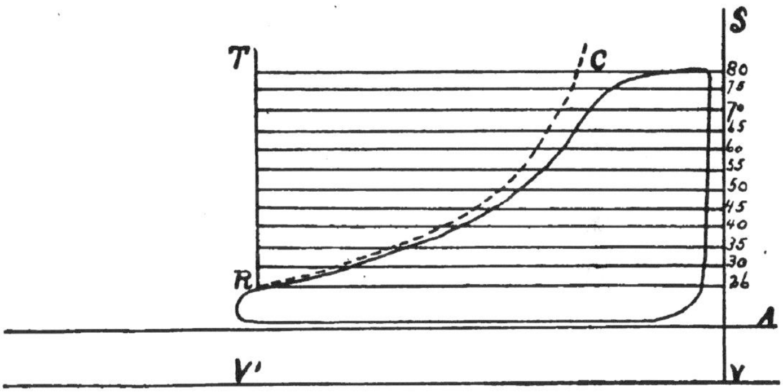

Fig. 187.

The dotted line R C shows what the true adiabatic curve would be on the diagram, provided it could be realized.

Ans.—If it were possible to so protect or insulate the cylinder of a steam engine that there would be absolutely no transmission of heat either to or from the steam during expansion, a true adiabatic curve or "curve of no transmission" might be obtained. The closer the actual expansion curve of a diagram conforms to such a curve, the higher will be the efficiency of the engine as a machine for converting heat into work.

CHAPTER X

THE STEAM TURBINE—FUNDAMENTAL PRINCIPLES

Ques. 707.—What are the basic principles governing the action of steam turbines?

Ans.—There are two fundamental principles upon which all steam turbines operate, viz., reaction and impulse. In some types of turbines the reaction principle alone is utilized, and in others the impulse, while in still others, and probably the most successful ones, both principles are combined.

Ques. 708.—In what general direction does the steam flow when used in a turbine?

Ans.—Parallel with the shaft or rotor, and also in a screw-like direction around it. This definition does not apply, however, to turbines of the purely impulse type, like the De Laval, for instance.

Ques. 709.—What causes the rotor to revolve?

Ans.—The action of the steam, coming, as it does, with tremendous velocity and great force against the small buckets or vanes with which the rotor is fitted, causes it to revolve, and as there is a continuous current of steam passing into the cylinder, the motion is continuous.

Ques. 710.—What law of turbo-mechanics governs the relation of bucket-speed, and fluid or steam speed?

Ans.—For purely impulse-wheels, bucket-speed equals one-half of jet-speed. For reaction wheels, bucket-speed equals jet-speed.

Ques. 711.—With what velocity would steam of 100 pounds pressure discharge into a vacuum of 28 inches?

Ans.—The theoretical velocity would be 3,860 feet per second.

Ques. 712.—What amount of energy would a cubic foot of steam under 100 pounds pressure exert if allowed to discharge into a vacuum of 28 inches?

Ans.—59,900 foot pounds.

Ques. 713.—Does the steam impinge against the first rows or sections of buckets at full pressure?

Ans.—In turbines of the Parsons type, the initial pressure of the steam is practically boiler-pressure, but it gradually falls as it passes on through the cylinder, which becomes larger in diameter as the exhaust end is approached. In other types of turbines, the steam is admitted to and directed against the blades or buckets, through expanding nozzles, and by the time it strikes the first stage, or section of moving vanes, the pressure has fallen to one-third or less of the original boiler-pressure, but the velocity is very great.

Ques. 714.—In what particular respect does the steam turbine appear to possess an advantage over the reciprocating engine, in the use of steam?

Ans.—The turbine, if designed along correct lines, is capable of utilizing in the highest degree one of the most valuable properties of steam, viz., velocity.

Ques. 715.—Give an example of the great increase in the amount of work performed by an agent when velocity is one of the factors made use of.

Ans.—Suppose that a man is standing within arm's

length of a heavy plate-glass window and that he holds in his hand an iron ball weighing 10 pounds. Suppose the man should place the ball against the glass and press the same there with all the energy he is capable of exerting. He would make very little, if any, impression upon the glass. But suppose that he should walk away from the window a distance of 20 feet, and then exert the same amount of energy in throwing the ball against the glass, a different result would ensue. The velocity with which the ball would impinge against the surface of the glass would no doubt ruin the window. Now, notwithstanding the fact that weight, energy, and time involved were exactly the same in both instances, yet a much larger amount of work was performed in the latter case, owing to the added force imparted to the ball by the velocity with which it impinged against the glass.

Ques. 716.—Describe the construction and action of the De Laval steam turbine.

Ans.—The De Laval steam turbine is termed by its builders a high-speed rotary steam-engine. It has but a single wheel, fitted with vanes or buckets of such curvature as has been found to be best adapted for receiving the impulse of the steam-jet. There are no stationary or guide-blades, the angular position of the nozzles giving direction to the jet. The nozzles are placed at an angle of 20 degrees to the plane of motion of the buckets. The heat energy in the steam is practically devoted to the production of velocity in the expanding or divergent nozzle, and the velocity thus attained by the issuing jet of steam is about 4,000 feet per second. To attain the maximum of efficiency, the buckets attached to the periphery of the wheel against which this jet impinges should have a speed of about 1,900 feet per second, but, owing to the difficulty of producing a material for the wheel strong enough to withstand the strains induced by

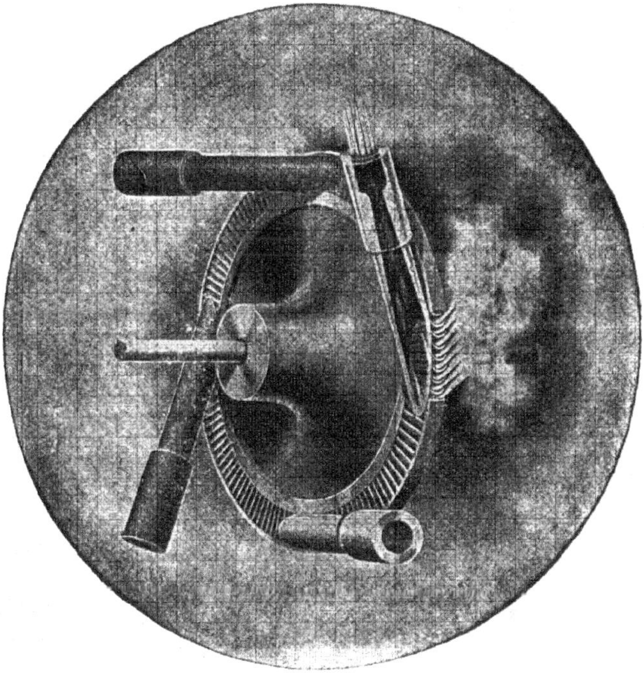

FIG. 188. THE DE LAVAL TURBINE WHEEL AND NOZZLES.

such a high speed, it has been found necessary to limit the peripheral speed to 1,200 or 1,300 feet per second.

Ques. 717.—Describe the action of the steam in its passage through the De Laval diverging nozzle.

Ans.—It is well known that in a correctly designed nozzle the adiabatic expansion of the steam from maximum to minimum pressure will convert the entire static energy of the steam into kinetic. Theoretically this is what occurs in the De Laval nozzle. The expanding steam acquires great velocity, and the energy of the jet of steam

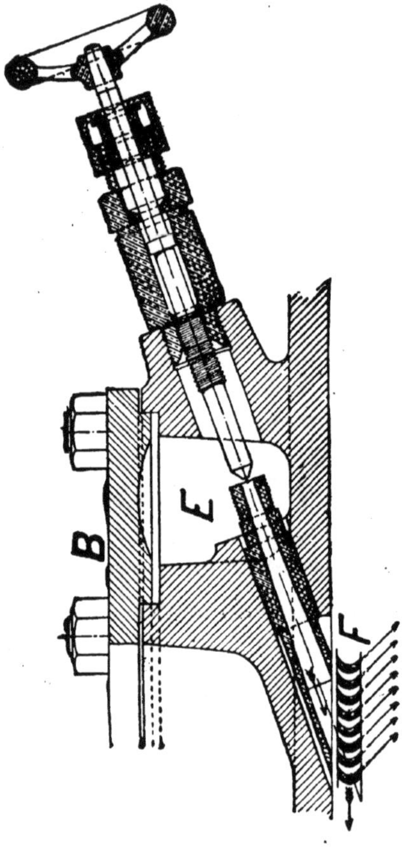

FIG. 189. DE LAVAL NOZZLE, SECTIONAL VIEW.

issuing from the nozzle is equal to the amount of energy that would be developed if an equal volume of steam were allowed to adiabatically expand behind the piston of a reciprocating engine, a condition, however, which for obvious reasons has never yet been attained in practice with the reciprocating engine. But with the divergent nozzle the conditions are different.

Ques. 718.—What is the usual speed of the De Laval steam-turbine wheel?

Ans.—From 10,000 to 30,000 revolutions per minute, according to the size of the machine.

Ques. 719.—How are the difficulties attending such high velocities overcome?

Ans.—By the long, flexible shaft and the ball and socket type of bearings, which allow of a slight flexure of the shaft in order that the wheel may revolve about its center of gravity rather than the geometrical center or center of position. All high-speed parts of the machine are made of forged nickel steel of great tensile strength.

Ques. 720.—How is the speed of the De Laval turbine-wheel and shaft reduced and transmitted for practical purposes?

Ans.—By a pair of very perfectly cut spiral gears, usually made 10 to 1. These gear-wheels are made of solid cast steel, or of cast iron with steel rims pressed on. The teeth in two rows are set at an angle of 90 degrees to each other. This arrangement insures smooth running and at the same time checks any tendency of the shaft towards end-thrust, thus dispensing with a thrust bearing.

Ques. 721.—How are the buckets made and fitted to the De Laval wheel?

Ans.—The buckets are drop-forged and made with a bulb shank, fitted in slots, that are milled in the rim of the wheel.

Ques. 722.—How many buckets are there?

Ans.—The number of buckets varies according to the capacity of the machine. There are about 350 buckets

Fig. 190 shows a 110 H. P. De Laval turbine and rotary pump with the upper half of the gear case and field frame removed for purposes of inspection. The slender shaft is seen projecting from the center of the turbine case, and upon this shaft are shown the small pinions meshing into the large spiral gears upon the two pump shafts.

on a 300 horse-power wheel, which is the largest size built up to the present time.

Ques. 723.—How many of the diverging nozzles are fitted to each wheel?

Ans.—The number of these nozzles depends upon the size of the machine, ranging from one to fifteen. They are generally fitted with shut-off valves by which one or more nozzles can be cut out when the load is light. This

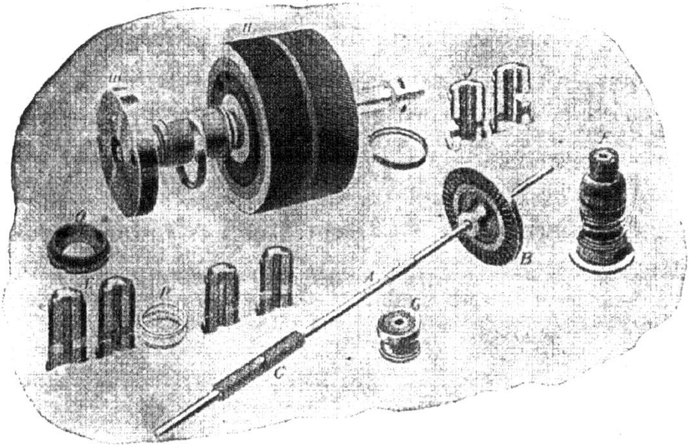

FIG. 191. WORKING PARTS OF THE DE LAVAL STEAM TURBINE.

A.—Turbine shaft.
B.—Turbine wheel.
C.—Pinion.
D.—Pinion bearing, two parts.
E.—Pinion bearing, two parts.
F.—Wheel bearing with spring.
G.—Flexible bearing.
H.—Gear wheel.
I.—Gear wheel shaft.
J.—Gear wheel bearing, two parts.
K.—Oil ring.
L.—Gear wheel bearing in position.
M.—Coupling.
N.—Centrifugal governor.
O.—Gland adjusting nut.
P.—Adjusting nut for flexible bearing.

renders it possible to use steam at boiler-pressure, no matter how small the volume required for the load. This is a matter of great importance, especially where the load varies considerably, as, for instance, there are plants in which during certain hours of the day a 300 horse-power machine may be taxed to its utmost capacity and during certain other hours the load on the same machine may drop to 50 horse-power. In such cases the number of nozzles in action may be reduced by closing the shut-off valves until the required volume of steam is admitted to the wheel. This adds to the economy of the machine. After passing through the nozzles, the steam, as elsewhere explained, is now completely expanded, and in impinging on the buckets its kinetic energy is transferred to the turbine wheel. Leaving the buckets, the steam now passes into the exhaust-chamber, and out through the exhaust-opening, to the condenser or atmosphere, as the case may be.

Ques. 724.—How is the speed of this turbine regulated?

Ans.—The governor is of the centrifugal type, although differing greatly in detail from the ordinary fly-ball governor. It is connected directly to the end of the gear-wheel shaft.

Ques. 725.—Describe the methods of lubricating the bearings on the De Laval turbine.

Ans.—The main shaft and dynamo bearings are ring-oiling. The high-speed bearings on the turbine shaft are fed by gravity from an oil-reservoir, and the drip-oil is collected in the base and may be filtered and used again.

Ques. 726.—What can be said regarding the steam-consumption of this turbine?

Ans.—Efficiency tests of the De Laval turbine show a high economy in steam-consumption, as, for instance, a test made by Messrs. Dean and Main, of Boston, Mass., on a 300 horse-power turbine, using saturated steam at about 200 pounds pressure per square inch and developing 333 brake horse-power, showed a steam-consumption of 15.17 pounds per brake horse-power, and the same machine, when supplied with superheated steam and carrying a load of 352 brake horse-power, consumed but

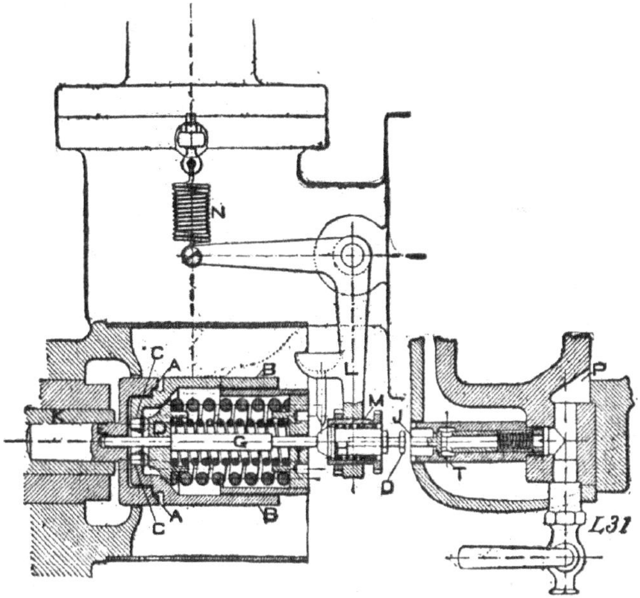

FIG. 192. THE DE LAVAL STEAM TURBINE GOVERNOR.

Two weights B are pivoted on knife edges A with hardened pins C, bearing on the spring seat D. E is the governor body fitted in the end of the gear wheel shaft K and has seats milled for the knife edges A. It is afterwards reduced in diameter to pass inside of the weights and its outer end is threaded to receive the adjusting nut I, by means of which the tension of the spring, and through this the speed of the turbine, is adjusted. When the speed accelerates, the weights, affected by centrifugal force, tend to spread apart, and pressing on the spring seat at D push the governor pin G to the right, thus actuating the bell crank L and cutting off a part of the flow of steam.

13.94 pounds per brake horse-power. These results compare most favorably with those of the highest type of reciprocating engines.

Ques. 727.—Since the steam is used in but a single

stage or section of buckets in the De Laval turbine, why such good economy in the use of steam?

Ans.—The static energy in the steam as it enters the nozzles is converted into kinetic energy by its passage through the divergent nozzles, and the result is a greatly increased volume of steam leaving the nozzles at a tremendous velocity, but at a greatly reduced pressure—practically exhaust pressure—impinging against the buckets of the turbine wheel and thus causing it to revolve.

TABLE No. 11

CAPACITIES AND SPEED OF DE LAVAL TURBINES

Horse Power.	Revolutions Turbine Shaft.	Revolutions Main Shaft.	Approximate Weight, Pounds.
5	30,000	3,000	330
10	24,000	2,400	650
20	20,000	2,000	1,250
75	16,400	1,500	5,000
110	13,000	1,200	8,000
225	11,060	900	15,000
300	10,500	900	20,000

Ques. 728.—Describe in general terms the Curtis steam-turbine.

Ans.—The Curtis turbine is built by the General Electric Company at their works in Schenectady, N. Y., and Lynn, Mass. The larger sizes are of the vertical type, and those of small capacity are horizontal. In the vertical type the revolving parts are set upon a vertical shaft, the diameter of the shaft corresponding to the size of the machine. The shaft is supported by and runs upon a step-bearing at the bottom. This step-bearing consists of two cylindrical cast-iron plates bearing upon each other and having a central recess between them into which lubricating oil is forced under pressure by a steam or electrically driven pump, the oil passing up from

FIG. 193. 5,000 K. W. CURTIS STEAM TURBINE DIRECT CONNECTED TO 5,000 K. W. THREE-PHASE ALTERNATING CURRENT GENERATOR.

beneath. A weighted accumulator is sometimes installed in connection with the oil pipe as a convenient device for governing the step-bearing pumps, and also as a safety device in case the pumps should fail, but it is seldom required for the latter purpose, as the step-bearing pumps have proven, after a long service in a number of cases, to be reliable. The vertical shaft is also held in place and kept steady by three sleeve bearings, one just above the step, one between the turbine and generator, and the other near the top. These guide bearings are lubricated by a standard gravity feed system. It is apparent that the amount of friction in the machine is very small, and as there is no end-thrust caused by the action of the steam, the relation between the revolving and stationary blades may be maintained accurately. As a consequence, therefore, the clearances are reduced to the minimum. The Curtis turbine is divided into two or more stages, and each stage has one, two or more sets of revolving blades bolted upon the peripheries of wheels keyed to the shaft. There are also the corresponding sets of stationary blades, bolted to the inner walls of the cylinder or casing.

Ques. 729.—What is the diameter of the vertical shaft for a 5,000 kilowatt turbine and dynamo?

Ans.—Fourteen inches.

Ques. 730.—How is the heat energy in the steam imparted to the wheel of the Curtis turbine?

Ans.—Both by impulse and reaction. The steam is admitted through expanding nozzles in which nearly all of the expansive force of the steam is transformed into the force of velocity. The steam is caused to pass through one, two, or more stages of moving elements, each stage having its own set of expanding nozzles, each succeeding set of nozzles being greater in number and of larger area than the preceding set. The ratio of expansion within these nozzles depends upon the number of stages, as, for instance, in a two-stage machine the steam enters the initial set of nozzles at boiler-pressure, say 180 pounds. It leaves these nozzles and enters the first set of moving blades at a pressure of about 15 pounds, from which it further expands to atmospheric pressure in passing

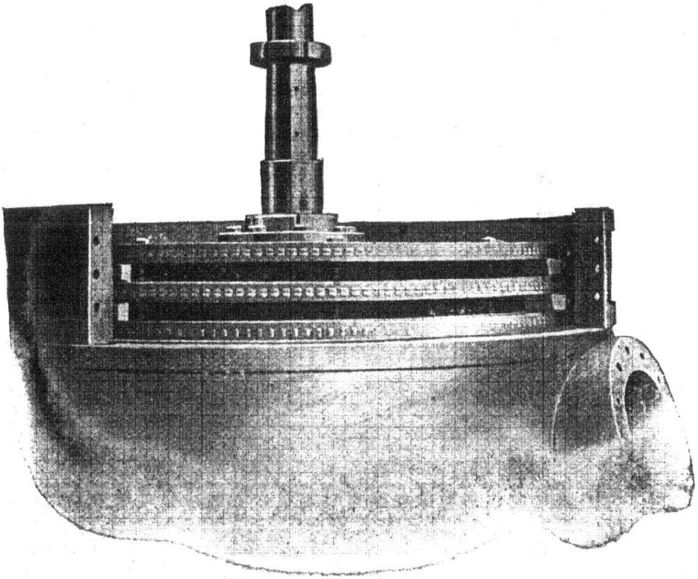

FIG. 194. ONE STAGE OF A 500 K. W. CURTIS STEAM TURBINE IN COURSE OF CONSTRUCTION.

through the wheels and intermediates. From the pressure in the first stage the steam again expands through the larger area of the second stage nozzles to a pressure slightly greater than the condenser vacuum at the entrance to the second set of moving blades, against which it now impinges and passes through, still doing work, due to velocity and mass. From this stage the steam passes to the condenser. If the turbine is a four-stage machine and the initial pressure is 180 pounds, the pressure at the different stages would be distributed in

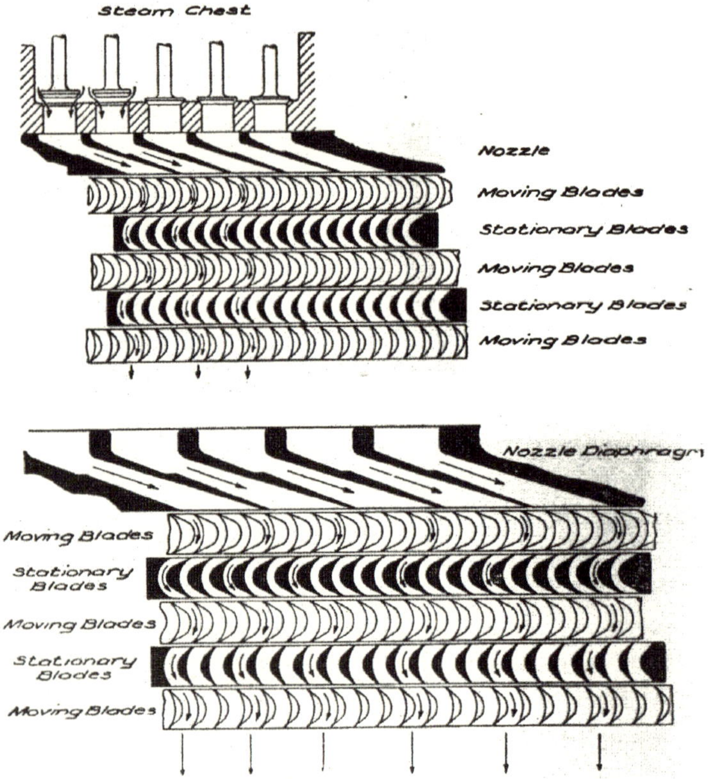

FIG. 195.

Diagram of the nozzles, moving blades and stationary blades of a two-stag Curtis steam turbine. The steam enters the nozzle openings at the top, controlled by the valves shown, two of the valves are open, and the course of the steam through the first stage is indicated by the arrows.

about the following manner: Initial pressure, 180 pounds; first stage, 50 pounds; second stage, 5 pounds;

third stage, partial vacuum, and fourth stage, condenser vacuum.

Ques. 731.—What are the diameters of the wheels?

Ans.—The diameters of the wheels vary according to the size of the machine, that of a 5,000 kilowatt unit being 13 feet.

Ques. 732.—What amount of clearance is there between the revolving and stationary blades?

Ans.—The clearance between the revolving and stationary blades is from $\frac{1}{32}$ to $\frac{1}{16}$ inch, thus reducing the wastage of steam to a very low percentage.

Ques. 733.—Describe the action of the steam in a two-stage Curtis turbine.

Ans.—The steam enters the nozzle openings at the top through valves that are controlled by the governor. After passing successively through the different sets of moving blades and stationary blades in the first stage, the steam passes into the second steam-chest. The flow of steam from this chamber to the second stage of buckets is also controlled by valves, but the function of these valves is not in the line of speed-regulation, but for the purpose of limiting the pressure in the stage-chambers, in a manner somewhat similar to the control of the receiver pressure in a two-cylinder or three-cylinder compound reciprocating engine. The valves controlling the admission of steam to the second and later stages differ from those in the first group in that they partake more of the nature of slide-valves and may be operated either by hand or automatically; in fact, they require but very little regulation, as the governing is always done by the live-steam admission-valves. As previously stated, the steam first strikes the moving blades in the first stage of a two-stage machine at a pressure of about 15 pounds

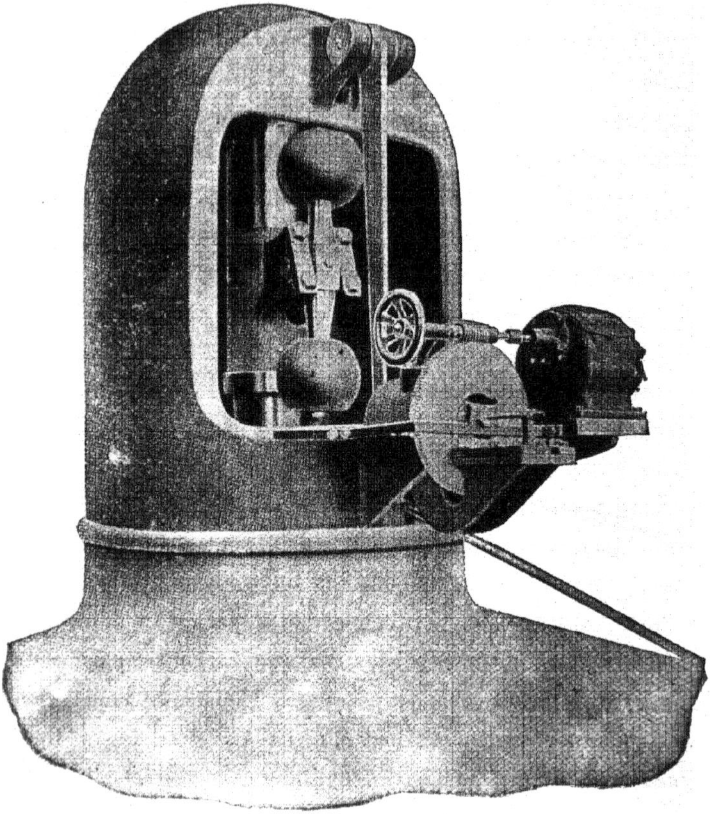

FIG. 196. GOVERNOR FOR 5,000 K. W. TURBINE.

above atmospheric pressure, but with great velocity. From this wheel it passes to the set of stationary blades between it and the next lower wheel. These stationary blades change the direction of flow of the steam and cause it to impinge against the buckets of the second wheel at the proper angle.

Ques. 734.—How is speed-regulation accomplished in the Curtis steam turbine?

Ans.—The governing of speed is accomplished in the first set of nozzles, and the control of the admission-valves here is effected by means of a centrifugal governor attached to the top end of the shaft. This governor, by

FIG. 197. ELECTRICALLY OPERATED VALVE.

a very slight movement, imparts motion to levers, which in turn work the valve mechanism. The admission of steam to the nozzles is controlled by piston-valves which are actuated by steam from small pilot-valves which are in turn under the control of the governor. Speed-regulation is effected by varying the number of nozzles in flow, that is, for light loads fewer nozzles are open and a smaller volume of steam is admitted to the turbine wheel, but the steam that is admitted impinges against the moving blades with the same velocity always, no matter whether the volume be large or small. With a full load and all the nozzle sections in flow, the steam passes to the wheel in a broad belt and steady flow.

FIG. 198. 5,000 KILOWATT GENERATING UNITS.

Comparison of space occupied and size of foundations. Modern Engine Type Unit and a Westinghouse-Parsons Turbine Type Unit of similar rating and overload capacity.

Ques. 735.—What great advantage does the steam-turbine as a prime mover for an electric generator possess over the reciprocating engine?

Ans.—The advantage of a high speed of revolution,

whereby there can be a great reduction in the size, weight, and cost of the direct-driven generator.

Ques. 736.—Give approximately the over-all dimensions of a Westinghouse-Parsons turbo-generator unit of 5,500-kilowatt, 11,000 volt capacity, of the revolving field type, speed 750 revolutions per minute, vacuum to be 27½ inches.

Ans.—Length 47 feet, width 13 feet, and height 14 feet to top of gallery-ring.

FIG. 199. GENERAL VIEW OF A 400 K. W. TURBINE GENERATOR UNIT.

Ques. 737.—What amount of floor-space would a reciprocating engine and direct-connected generator of equal capacity with the above occupy?

Ans.—The generator would be 42 feet in extreme diameter, its weight would be 445 tons (speed to be 75 revolutions per minute) and it, together with the four-cylinder piston engine, would fill a space 40 feet wide by 60 feet long, and tower 45 feet in height.

Ques. 738.—Describe in general terms the construction and principles of operation of the Westinghouse-Parsons steam-turbine.

Ans.—The Westinghouse-Parsons steam-turbine is fundamentally based upon the invention of Mr. Charles A. Parsons, who, while experimenting with a reaction turbine constructed along the lines of Hero's engine, conceived the idea of combining the two principles, reaction

FIG. 200. Shows a 600 H. P. machine with the upper half of the cylinder, or stator as it is termed, thrown back for inspection.

and impulse, and also of causing the steam to flow in a general direction parallel with the shaft of the turbine. This principle of parallel flow is common to all four types of turbines, but is perhaps more prominent in the Westinghouse-Parsons and less so in the De Laval. The cylinder, or stator, as it is termed, is divided longitudinally into an upper and a lower half flanged and bolted together. There are three sections or drums, gradually increasing in diameter from the inlet to the third and last group of blades. This arrangement may be likened in some measure to the triple-compound reciprocating engine.

Ques. 739.—Describe the arrangement of the blades or buckets in the Westinghouse-Parsons steam-turbine.

Ans.—There are two kinds of blades, viz., stationary blades and moving blades, but they are similar in shape, being of the same curvature. These blades are made of hard drawn material, and are set into their places and secured by a caulking process. The stationary blades project from the inside surface of the cylinder, while similar rows of moving blades project from the surface of the rotor, or revolving drum. When the upper half of the cylinder is in position each row of stationary blades fits in between two corresponding rows of moving blades.

Ques. 740.—Are these blades all of the same length?

Ans.—They are not. The length varies from ½ inch for the shortest to 7 inches for the longest, according to their location. The shortest blades are placed at the steam end of each section and the longest blades are placed at the opposite end.

Ques. 741.—What is the clearance between the blades as they stand in the rows?

Ans.—The clearance between the blades as they stand in the rows is ⅛ inch for the smallest size blades and ½ inch for the larger ones, gradually increasing from the inlet to the exhaust. In the 5,000 kilowatt machine the clearance at the exhaust end between the rows of blades is 1 inch.

Ques. 742.—What is the general direction taken by

the steam in its passage through the Westinghouse-Parsons turbine?

Ans.—The steam entering at the smaller end of the cylinder presses first against the shortest blades and then passes on through in the form of spiral or screw line about the rotor, continually pressing against new and gradually lengthening blades, thus doing work by reason of its velocity.

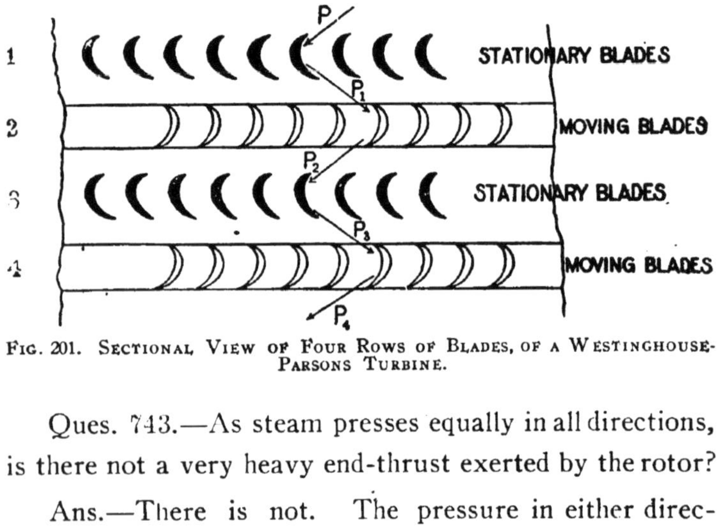

FIG. 201. SECTIONAL VIEW OF FOUR ROWS OF BLADES, OF A WESTINGHOUSE-PARSONS TURBINE.

Ques. 743.—As steam presses equally in all directions, is there not a very heavy end-thrust exerted by the rotor?

Ans.—There is not. The pressure in either direction is perfectly balanced by means of balancing pistons placed on the steam end of the rotor. The diameters of these pistons correspond to the diameters of the different drums or sections.*

Ques. 744.—About what is the velocity of the steam in the Parsons turbine?

*The theory and action of these balancing pistons is fully and completely described in Swingle's "Twentieth Century Hand Book for Engineers and Electricians."

Ans.—The highest velocity does not exceed 600 feet a second.

Ques. 745.—About what amount of pressure is exerted upon each blade by the steam?

Ans.—The steam-thrust on each blade is said to be equal to about 1 ounce avoirdupois.

Ques. 746.—With such a very light pressure upon

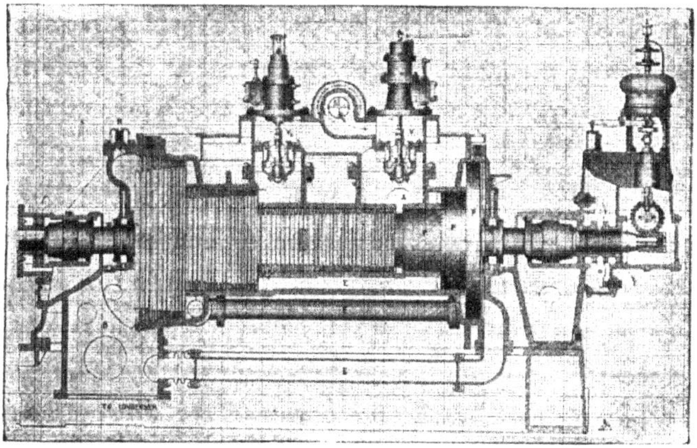

FIG. 202. SECTIONAL VIEW OF A WESTINGHOUSE-PARSONS TURBINE, SHOWING ARRANGEMENT OF BALANCING PISTONS P. P. P.

each blade, why is it that this turbine is capable of developing power?

Ans.—Because of the large number of blades; as, for instance, taking a 400 kilowatt machine, there are 16,095 moving blades and 14,978 stationary blades, a total of 31,073.

Ques. 747.—How are the clearances preserved?

Ans.—A rigid shaft and thrust or adjustment bearing accurately preserves the clearances.

Ques. 748.—Describe the construction and action of the bearings.

Ans.—The bearings are constructed along lines differing from those of the ordinary reciprocating engine. The bearing proper is a gun-metal sleeve that is prevented from turning by a loose-fitting dowell. Outside of this sleeve are three concentric tubes having a small clearance between them. This clearance is kept constantly filled with oil supplied under light pressure, which permits a vibration of the inner shell or sleeve and at the same time tends to restrain or cushion it. This arrangement allows the shaft to revolve about its axis of gravity instead of the geometrical axis, as would be the case if the bearing were of the ordinary construction. The journal is thus to a certain degree a floating journal, free to run slightly eccentric according as the shaft may happen to be out of balance.

Ques. 749.—How is the power of the Westinghouse-Parsons turbine transmitted to the dynamo, or other machine to be run?

Ans.—A flexible coupling is provided, by means of which the power of the turbine is transmitted to the dynamo or other machine it is intended to run. The oil from all the bearings drains back into a reservoir, and from there it is forced up into a chamber, where is forms a static head, which gives a constant pressure of oil on all the bearings.

Ques. 750.—How is the speed governed?

Ans.—The speed of the Westinghouse-Parsons turbine is regulated by a fly-ball governer constructed in such manner that a very slight movement of the balls serves to produce the required change in the supply of steam. The ball levers swing on knife edges instead of pins. The governor works both ways, that is to say, when the levers are oscillating about their mid position a head of steam corresponding to full load is being admitted to the turbine, and a movement from this point, either up or down, tends to increase or to decrease the supply of steam.

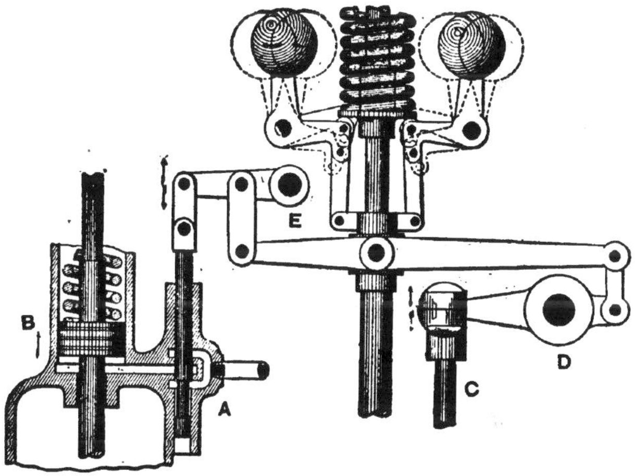

FIG. 203. SECTION OF WESTINGHOUSE-PARSONS TURBINE GOVERNOR.

Ques. 751.—What can be said of the efficiency of the Westinghouse-Parsons steam-turbine?

Ans.—Under test a 400 kilowatt Westinghouse-Parsons steam-turbine, using steam at 150 pounds initial pressure and superheated about 180 degrees, consumed 11.17 pounds of steam per brake horse-power hour at full load. The speed was 3,550 revolutions per minute and the vacuum was 28 inches. With dry saturated steam the consumption was 13.5 pounds per brake horse-power hour at full load, and 15.5 pounds at one-half load. A 1,000 kilowatt machine, using steam of 150 pounds pressure and superheated 140 degrees, exhausting into a vacuum of 28 inches, showed the very remarkable economy of 12.66 pounds of steam per electrical horse-power per hour. A 1,500 kilowatt Westinghouse-Parson turbine, using dry saturated steam of 150 pounds pressure with 27 inches vacuum, consumed 14.8 pounds steam per electrical horse-power hour at full load, and 17.2 pounds at one-half load.

Ques. 752.—What efficiency does the Curtis turbine show in the use of steam?

Ans.—A 600 kilowatt Curtis turbine operating at 1,500 revolutions per minute, with steam at 140 pounds gauge-pressure and 28.5 inches vacuum, showed a steam-consumption as follows, steam superheated 150 degrees: At full load, 12.5 pounds per electrical horse-power per hour; at half load, 13.25 pounds per electrical horse-power per hour; at one-sixth load, 16.2 pounds per electrical horse-power per hour, and at one-third overload, 12.4 pounds per electrical horse-power per hour.

Ques. 753.—Describe in brief terms the Hamilton-Holzwarth steam-turbine.

Ans.—The Hamilton-Holzwarth steam-turbine is based upon and has been developed from the designs of Prof. Rateau, and is being manufactured in this country by the Hooven-Owens-Rentschler Company, of Hamilton, Ohio. It is horizontal and placed upon a rigid bed-plate of the box pattern. All steam, oil and water-pipes are within and beneath this bed-plate, as are also the steam-inlet-valve and the regulating and by-pass valves. The smaller sizes of this turbine are built in a single casing or cylinder, but for units of 750 kilowatts and larger the revolving element is divided into two parts, high and low pressure. This turbine resembles the Westinghouse-Parsons turbine in some respects, prominent of which is that it is a full-stroke turbine, that is, that the steam flows through it in one continuous belt or veil in screw line, the general direction being parallel with the shaft. But, unlike the Parsons type, the steam in the Hamilton-Holzwarth turbine is made to do its work only by impulse, and not by impulse and reaction combined. It might thus be termed an action turbine.

Ques. 754.—Describe the interior construction of this turbine.

Ans.—The interior of the cylinder is divided into a series of stages by stationary disks which are set in grooves in the cylinder and are bored in the center to allow the shaft, or rather the hubs of the running wheels that are keyed to the shaft, to revolve in this bore. There are no balancing pistons in this machine, the axial thrust of the shaft being taken up by a thrust ball-bearing. Between each two stationary disks there is located a running wheel, and the clearance between the running vanes and the stationary vanes is made as slight as is consistent with safe practice.

Ques. 755.—Describe the construction of the running vanes and the action of the steam upon them.

Ans.—The running vanes conform in section somewhat

to the Parsons type, but the action of the steam upon them and also within the stationary vanes is different. The expansion of the steam and consequent development of velocity takes place entirely within the stationary vanes, which also change the direction of flow of the steam and distribute it in the proper manner to the vanes of the running wheels, which, according to the claims of the makers, the steam enters and leaves at the same pressure, thus allowing the wheel to revolve in a uniform pressure.

FIG. 204. GENERAL VIEW OF A HAMILTON-HOLZWARTH STEAM-TURBINE.

Ques. 756.—What provision is made in the Hamilton-Holzwarth turbine for maintaining the velocity of the steam as it expands?

Ans.—The first stationary disk of the low-pressure turbine has guide-vanes all around its circumference, so that the steam enters the turbine in a full cylindrical belt, interrupted only by the guide-vanes. To provide for the increasing volume as the steam expands in its course through the turbine, the areas of the passages through the distributers and running vanes must be progressively enlarged. The gradual increase in the dimensions of the stationary vanes permits the steam to expand within them, thus tending to maintain its velocity, while at the same time the vanes guide the steam under such a small angle that the force with which it impinges against the vanes of the next running wheel is as effective as possible. The curvature of the vanes is such that the steam while passing through them will increase its velocity in a ratio corresponding to its operation.

Ques. 757.—Describe the method of regulating the speed of this turbine.

Ans.—The governor is of the spring and weight type, adapted to high speed, and is designed especially for turbine governing. It is directly driven by the turbine-shaft, revolving with the same angular velocity. Its action is as follows: Two disks keyed to the shaft, drive, by means of rollers, two weights sliding along a cross-bar placed at right angles through the shaft and compressing two springs against two nuts on the cross-bar. Every movement of the weights, caused by increasing or decreasing the angular velocity of the turbine-shaft, is transmitted by means of levers to a sleeve which actuates the regulating mechanism. These levers are balanced so that no back pressure is exerted upon the weights. The whole governor is closed in by the disks, one on each side, and a steel ring secured by concentric recesses to the disks. In order to decrease the friction within the governor and regulating mechanism, thrust ball-bearings and frictionless roller-bearings are used.

Ques. 758.—Describe the action of the steam within the Hamilton-Holzwarth steam-turbine.

Ans.—After leaving the steam-separator that is located beneath the bed-plate, the steam passes through the inlet or throttle-valve, the stem of which extends up through the floor near the high-pressure casing and is protected by a floor-stand and equipped with a hand wheel. The steam now passes through the regulating valve, which will be described later on. From this valve it is led through a curved pipe to the front head of the high-pressure casing or cylinder. In this head is a ring channel into which the steam enters, and from whence it flows through the first set of stationary vanes.

Ques. 759.—Describe the action of the steam as it passes through the first set of stationary vanes.

Ans.—In these vanes the first stage of expansion occurs, the velocity of the flow is accelerated, and the direction of flow is changed by the curve of the vanes in such manner that the steam impinges against the vanes of the first running wheel at the proper angle and in a full cylindrical belt, imparting by impulse a portion of its energy to the wheel.

Ques. 760.—What takes place in the course of the steam after leaving the first running wheel?

Ans.—Passing through the vanes of this wheel, the steam immediately enters the vanes of the second stationary disk, which are larger in area than those of the first, and here occurs the second stage of expansion, another acceleration of velocity, and also the proper change in direction, and the steam leaves this distributer and impinges against the vanes of the second running wheel. This cycle is repeated throughout the several stages of the turbine, a certain percentage of the heat energy in the steam being imparted by impulse to each wheel and thence to the turbine-shaft. From the last running wheel the steam is led through receiver pipes to the front head of the low-pressure cylinder, or, if there is but one cylinder, directly to the condenser or the atmosphere.

Ques. 761.—Describe the construction and location of the regulating valve.

Ans.—The regulating valve is located beneath the bed-plate. One side of it is connected by a curved pipe with the front head of the high-pressure cylinder and the other side is connected with the inlet-valve. The regulating valve is of the double-seated poppet-valve type. Valves and valve-seats are made of tough cast steel, to avoid corrosion as much as possible, and the valve-body is made of cast iron.

Ques. 762.—Describe the by-pass regulating valve.

Ans.—This valve is also a double-seated poppet-valve and is located immediately below the regulating valve and forming a part of it. Thus the use of a second stuffing box for the stem of this valve is avoided. The function of this valve is to control the volume of the live-steam supply that flows directly to the by-pass nozzles in the front head of the low-pressure casing.

Ques. 763.—How is the main regulating valve operated?

Ans.—The main regulating valve is not actuated directly by the governor, but by means of the regulating mechanism.

Ques. 764.—Describe the construction and operation of the regulating mechanism of the Hamilton-Holzwarth steam-turbines.

Ans.—The construction and operation of this regulating mechanism is as follows: The stem of the regulating valve is driven by means of bevel gears by a shaft that is supported in frictionless roller-bearings. On this shaft there is a friction wheel that the governor can slide across the face of a continuously revolving friction disk by means of its sleeve and bell-crank lever. This revolving disk is keyed to a solid shaft which is driven by a coupling from a hollow shaft. This hollow shaft is driven by the turbine-shaft through the medium of a worm gear. The solid shaft, with the continuously revolving friction disk, can be slightly shifted by the governor sleeve so that the two friction disks come into contact when the sleeve moves, that is, when the angular velocity changes. If this change is relatively great, the sleeve will draw the periodically revolving friction disk far from the center of the always revolving one, and this disk will quickly drive the stem of the regulating valve and the flow of steam will thus be regulated. As soon as the angular velocity falls below a certain percentage of the normal speed, the driving friction disk is drawn back by the governor, the regulating valve remains open and the whole regulating mechanism rests or stops, although the shaft is still running.

Ques. 765.—Under what conditions will this governor shut down the turbine?

Ans.—Should the angular velocity of the shaft reach

a point 2.5 per cent higher than normal, the governor will shut down the turbine. If an accident should happen to the governor, due to imperfect material or breaking or weakening of the springs, the result would be a shut-down of the turbine.

Ques. 766.—How may the speed of this turbine be changed, while running, if necessary?

Ans.—In order to change the speed of the turbine while running, which might be necessary in order to run the machine parallel with another prime mover, a spring balance is provided, attached to the bell-crank lever of the regulating mechanism. The hand-wheel of this spring balance is outside of the pedestal for regulating mechanism and near the floor-stand and hand-wheel. With this spring balance the speed of the turbine may be changed 5 per cent either way from normal.

Ques. 767.—What is the best method of disposing of the exhaust steam of steam-turbines?

Ans.—As in the case of the reciprocating engine, the highest efficiency in the operation of the steam-turbine is obtained by allowing the exhaust steam to pass into a condenser, and experience has demonstrated that it is possible to maintain a higher vacuum in the condenser of a turbine than in that of a reciprocating engine. This is due, no doubt, to the fact that in the turbine the steam is expanded down to a much lower pressure than is possible with the reciprocating engine.

Ques. 768.—What type of condensing apparatus is best adapted to steam-turbines?

Ans.—The condensing apparatus used in connection

with steam-turbines may consist of any one of the modern improved systems, and as no cylinder-oil is used within the cylinder of the turbine, the water of condensation may be returned to the boilers as feed-water. If the condensing water is foul or contains matter that would be injurious to the boilers, a surface condenser should be used. If the water of condensation is not to be used in the boilers, the jet system may be employed.

Ques. 769.—What percentage of gain may be effected by allowing the exhaust steam from the turbine to pass into a good condenser?

Ans.—As an instance of the great gain in economy effected by the use of the condenser in connection with the steam-turbine, a 750 kilowatt Westinghouse-Parsons turbine, using steam of 150 pounds pressure, not superheated and exhausting into a vacuum of 28 inches, showed a steam consumption of 13.77 pounds per brake horse-power per hour, while the same machine operating non-condensing consumed 28.26 pounds of steam per brake horse-power hour. Practically the same percentage in economy effected by condensing the exhaust applies to the other types of steam-turbines.

Ques. 770.—About what is the additional cost of operating a complete condensing outfit in connection with a steam-turbine plant?

Ans.—With reference to the relative cost of operating the several auxiliaries necessary to a complete condensing outfit, the highest authorities on the subject place the power consumption of these auxiliaries at from 2 to 7 per cent of the total turbine output of power. A portion of this is regained by the use of an open heater for the feed-water, into which the exhaust steam from the auxiliaries may pass, thus heating the feed-water and returning a part of the heat to the boilers.

Ques. 771.—What precautions must be observed in the operation of a condensing outfit in order to obtain the highest efficiency?

Ans.—A prime requisite to the maintenance of high vacuum, with the resultant economy in the operation of the condensing apparatus, is that all entrained air must be excluded from the condenser. There are various ways in which it is possible for air to find its way into the condensing system. For instance, there may be an improperly packed gland, or there may be slight leaks in the piping, or the air may be introduced with the condensing water. This air should be removed before it reaches the condenser, and it may be accomplished by means of the "dry" air-pump.

Ques. 772.—Describe some of the leading characteristics of the dry air-pump.

Ans.—This dry air-pump is different from the ordinary air-pump that is used in connection with most condensing systems. The dry air-pump handles no water, the cylinder being lubricated with oil in the same manner as the steam-cylinder. The clearances also are made as small as possible. These pumps are built either in one or two stages.

Ques. 773.—What particular features would be required in the design of a compound or stage-expansion reciprocating engine, in order to develop a high vacuum, for instance as high as 28.5 inches?

Ans.—In comparing the efficiency of the reciprocating engine and the steam-turbine it is not to be inferred that reciprocating engines would not give better results at high vacuum than they do at the usual rate of 25 to 26 inches, but to reach and maintain the higher vacuum of 28 to 28.5 inches with the reciprocating engine would necessitate much larger sizes of the low-pressure cylinder, as also the valves and exhaust pipes, in order to handle the greatly increased volume of steam at the low-pressure demanded by high vacuum.

Ques. 774.—What advantage has the turbine over the reciprocating engine, in the disposal of its exhaust steam?

Ans.—The steam-turbine expands its working steam to within 1 inch of the vacuum existing in the condenser, that is, if there is a vacuum of 28 inches in the condenser there will be 27 inches of vacuum in the exhaust end of the turbine cylinder. On the other hand, there is usually a difference of 4 or 5 inches (2 to 2.5 pounds) between the mean back pressure in the cylinder of a reciprocating condensing engine and the absolute back pressure in the condenser.

Ques. 775.—Mention the two principal sources of economy that the steam-turbine possesses in a high degree.

Ans.—Two of the main sources of economy that the steam-turbine possesses in a much higher degree than does the reciprocating engine are: First, its adaptability for using superheated steam, and second, the possibility of maintaining a higher degree of vacuum.

Fig. 205. The Allis Chalmers Steam Turbine.

Ques. 776.—What can be said of the steam turbine, regarding friction of rubbing parts, such as reciprocating pistons, cross-heads, etc?

Ans.—There are no rubbing surfaces in the turbine except the bearings of the rotor.

Ques. 777.—Of what type is the Allis Chalmers steam-turbine?

Ans.—It is of the reaction, or Parsons type, with a number of modifications in details of construction.

Ques. 778.—Give an elementary description of the "Parsons" steam-turbine.

Ans.—It consists essentially of a fixed casing, or cylinder, usually arranged in three stages of different diameters, that of the smallest diameter being at the high-pressure, or admission end, and that of the largest diameter at the low-pressure or exhaust end of the casing.

Inside of this casing is a revolving drum, or rotor, the ends of which are extended in the form of a shaft, and carried in two bearings, just outside each end of the cylinder.

Ques. 779.—What causes the drum to revolve within the cylinder?

Ans.—The drum is fitted with a large number of small curved blades, or paddles arranged in straight rows around its circumference. The blades in each stage, or step, are also arranged in groups of increasing length, those at the beginning of each larger stage being shorter than those at the end of the preceding stage, the change being made in such a manner that the correct relation of blade length to drum diameter is secured. These rows of revolving blades fit in and run between corresponding rows of stationary blades that project from the walls of the cylinder. These stationary blades have the same curvature as the revolving blades, but are set so that the curves incline in the opposite direction to those of the revolving blades. The steam entering the cylinder at the smallest or first stage, is deflected in its course by the first row of stationary blades, and immediately impinges with a pressure but slightly reduced from boiler pressure, against the first row of revolving blades. It then passes

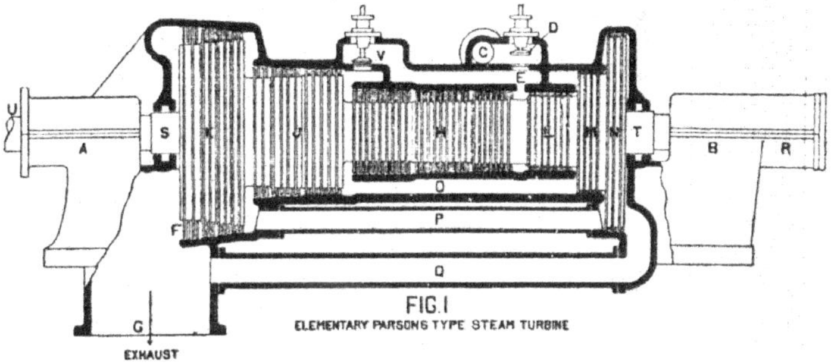

FIG. 206.

Main bearings, A and B. Thrust bearing, R. Steam pipe, C. Main throttle valve, D, which is balanced, and operated by the governor. Steam enters the cylinder through passage E, passes to the left through the alternate rows of stationary and revolving blades, leaving the cylinder at F and passes into the condenser, or atmosphere through passage G. H, J and K are the three steps or stages of the machine. L, M and N are the three balance pistons. O, P and Q are the equalizing passages, connecting the balance pistons with the corresponding stages.

to the next row of stationary blades, which again deflect its course so as to cause it to strike the next row of moving blades at the proper angle. Thus the continual pressure and reaction of the steam against the curved surfaces of the moving blades causes the drum, or rotor to revolve.

FIG. 207. SPINDLE OR ROTOR, ALLIS CHALMERS STEAM TURBINE.
The rings which carry the blades are pressed on.

Ques. 780.—Does not the action of the steam against the revolving blades tend to produce a strong end thrust?

Ans.—It does—but this thrust is neutralized by three "balance-pistons" so called, which are fitted upon the revolving drum at the high-pressure end of the cylinder. The diameter of each "piston" corresponds with the diameter of that stage of the cylinder with which it is connected by an equalizing passage which permits the steam to act upon it, and thus balance the thrust.

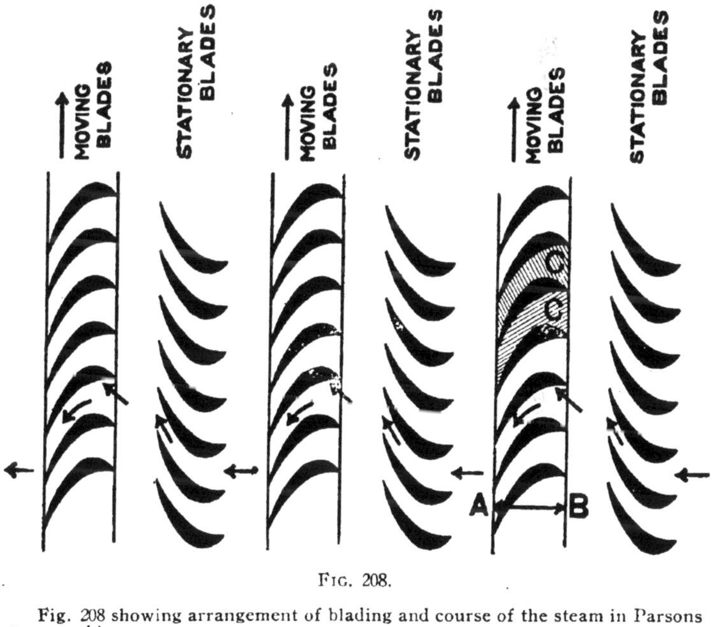

Fig. 208.

Fig. 208 showing arrangement of blading and course of the steam in Parsons steam turbine.

Ques. 781.—Do the revolving blades come in contact with the stationary parts?

Ans.—They do not. The high speeds which are necessary in the steam turbine prohibit any continuous contact between moving and stationary parts, except in the lubricated bearings.

Ques. 782.—How much clearance is allowed between the moving and stationary parts in the "Parsons" steam-turbine?

Ans.—The tips of the revolving blades just clear the walls of the cylinder, and the tips of the stationary blades just clear the surface of the rotor.

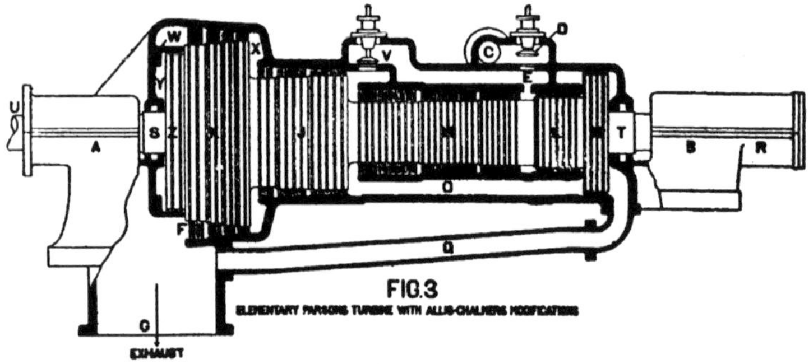

Fig. 209.

Sectional view of elementary Parsons steam turbine, with Allis Chalmers modifications. L and M are the two balance pistons at the high pressure end. Z is a smaller balance piston placed in the low pressure end, yet having the same effective area as did the larger piston N shown in Fig. 206. O and Q are the two equalizing passages for pistons L and M. Passage P is omitted in this construction and balance piston Z is equalized with the third stage pressure at Y. Valve V is a by-pass valve to allow of live steam being admitted to the second stage of the cylinder in case of a sudden overload. This by-pass valve is the equivalent of the by-pass valve used to admit live steam to the low pressure cylinder of a compound reciprocating engine. Valve V is arranged to be operated, either by the governor or by hand, as the conditions may require. Frictionless glands made tight by water packing are provided at S and T where the shaft passes out of the cylinder. The shaft is extended at U and connected to the generator shaft by a flexible coupling.

Ques. 783.—How are the clearances between the edges of the revolving and stationary blades preserved?

Ans.—The position of the drum, as regards end play, is definitely fixed by means of a small "thrust bearing" provided inside the housing of the main bearing.

This so-called thrust bearing can be adjusted to locate,

and hold the revolving spindle or rotor in such position as will allow sufficient clearance between the moving and stationary blades, and yet reduce the leakage of steam to a minimum.

Ques. 784.—Is there not danger of out leakage of steam, and in leakage of air, where the shaft passes out of the high and low-pressure ends of the cylinder?

Ans.—There is; but this is provided for by glands that are made practically frictionless by water packing, without metallic contact.

Ques. 785.—How is the power of the "Parsons" type of steam-turbine transmitted to the electric generator, or other machine to be run?

Ans.—The shaft is extended at the low-pressure end, and coupled to the shaft of the generator by means of a flexible coupling.

Ques. 786.—What provision is made in this type of steam-turbine for speed regulation?

Ans.—The speed of the "Parsons" turbine is regulated by a very sensitive governor driven from the turbine shaft by means of cut gears working in an oil bath. The governor operates a balanced throttle-valve, and may be adjusted for speed while the turbine is in motion if necessary for the synchronizing of alternators, and dividing the load.

Ques. 787.—Suppose there should be an accidental derangement of the governing mechanism, what provision is made for preventing dangerous over speed?

Ans.—A separate safety governor is provided, driven directly by the turbine shaft, without the intervention of gearing, and so adjusted that if the speed of the turbine should reach a predetermined point above that for which the main governor is set, the safety governor will come into action, and trip a valve, thus shutting off the steam, and stopping the turbine.

Ques. 788.—Is the arrangement of "balance-pistons" described in answer to question 780 carried out in all sizes of steam-turbines of the "Parsons" type?

Ans.—No. In the larger sizes of the Allis Chalmers steam-turbine, the largest one of the three pistons at the high-pressure end is replaced by a smaller balance-piston located at the low-pressure end of the turbine, and working inside a supplementary cylinder.

This piston presents the same effective area for the steam to act upon, as did the larger piston, because the working area of the latter in its original location consisted only of the annular area included between its periphery, and the periphery of the next smaller piston.

Ques. 789.—How is the pressure of the steam brought to bear upon this equalizing piston in its new position?

Ans.—By means of passages through the body of the rotor, connecting the third stage of the cylinder with the supplementary cylinder in which the piston revolves.

Ques. 790.—How are the blades or paddles fitted to, and held in the rotor, and cylinder of the Allis Chalmers steam-turbine?

Ans.—Each blade is individually formed by special machine tools, so that its root or foot is of an angular dove-tail shape, and at its tip there is a projection.

Foundation rings are provided for each row of blades.

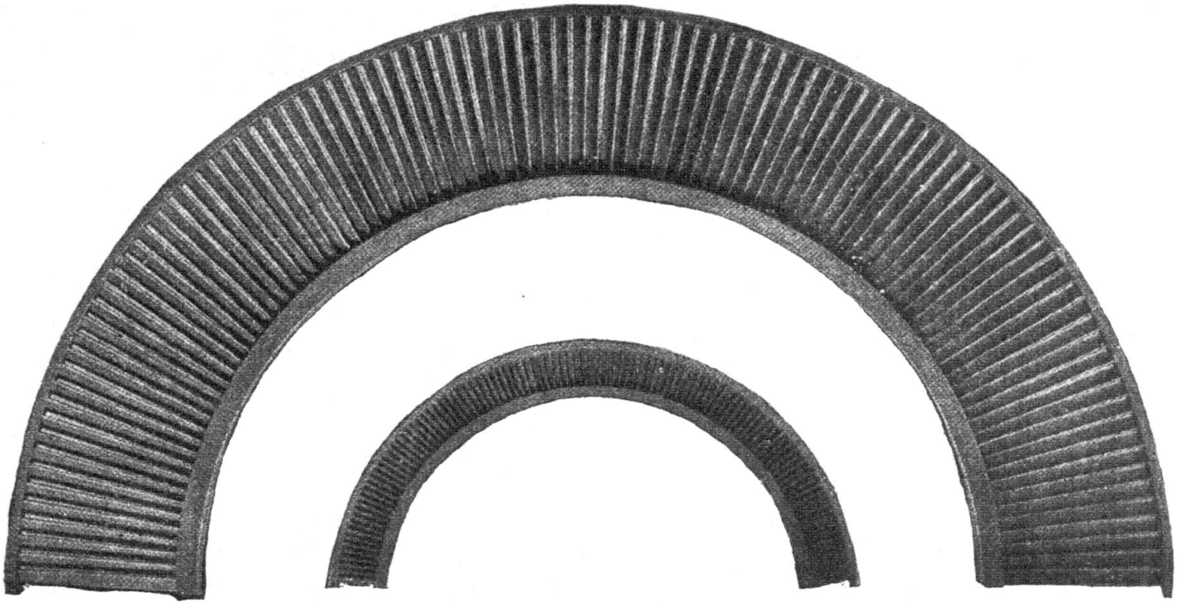

FIG. 210.

Half ring of blades inserted in the foundation ring before being placed upon the rotor, showing substantial construction.

These rings have slots of dove-tail shape cut into them to receive the roots of the blades. These slots are accurately spaced, and inclined so as to give the required pitch and angle to the blades. The foundation rings themselves are dove-tail in cross section, and are inserted in dove-tail grooves cut in the turbine cylinder, and rotor respectively. These rings are firmly held in place by key pieces that are driven into place, and upset into undercut grooves, thus locking the whole structure firmly together.

Ques. 791.—How are the tips or outer ends of the blades protected?

Ans.—By a shroud ring for each row, in which holes are punched to receive the projections on the tips of the blades.

These holes are spaced by special machinery to match the slots in the foundation ring.

Ques. 792.—Describe the construction of the shroud rings.

Ans.—They are channel shaped in cross section, and are made thin, so that in case of accidental contact with an opposing surface no dangerous heating will occur, neither will the rubbing be so liable to rip out the blades, as it is when they are unprotected by a shroud ring.

Ques. 793.—Mention another advantage in connection with the use of a shroud ring.

Ans.—The blades in each row are stiffened, and held together as a unit by its use, thus permitting smaller clearances, and reducing the leakage loss to a minimum. The channel shape of the shroud ring also forms an effective baffle to the steam leakage.

Fig. 211.

Fig. 211 illustrates blades as fitted in the rotor of Allis Chalmers steam turbine. The shroud ring protecting the tips of the blades is also shown.

Ques. 794.—What type of bearings are the Allis Chalmers steam-turbines fitted with?

Ans.—Self-adjusting ball and socket bearings especially designed for high speed, shims being provided for proper alignment.

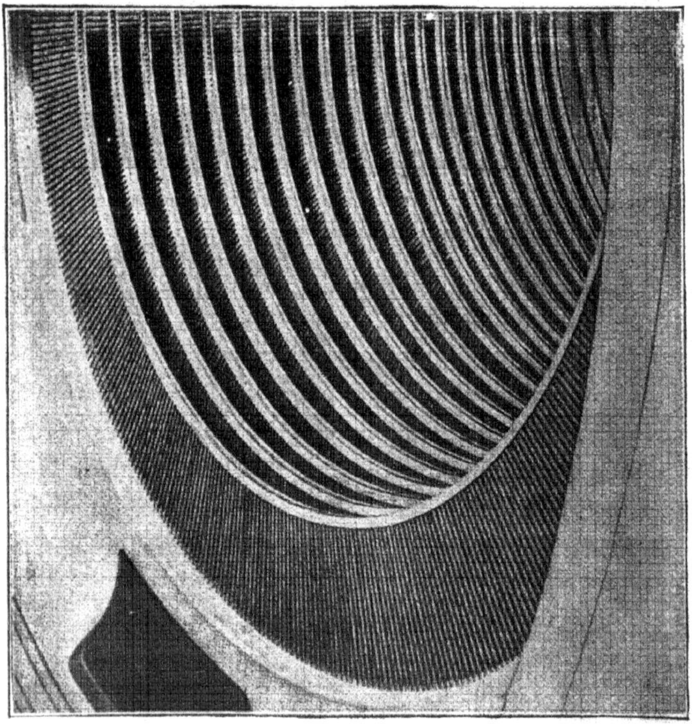

Fig. 212.

Fig. 212 shows a number of rows of stationary blades fitted in the cylinder of an Allis Chalmers steam turbine.

In the smaller sizes the bearing shells are made of special bronze, and in the larger sizes white metal is used for bearing surface.

Ques. 795.—How are these bearings lubricated?

Ans.—The oil is supplied freely to the middle of each bearing, and allowed to flow out at the ends, where it is caught, passed through a cooler, and pumped back to the bearings, to be used again and again.

Ques. 796.—Does the fact that the oil is supplied to the bearings in large quantities necessarily imply a heavy expenditure for oil?

Ans.—It does not; for the reason that the bearings practically float on oil films, thus preventing that "wearing out" of the oil which occurs when it is supplied in diminutive doses.

Ques. 797.—Can superheated steam be used to advantage in steam-turbines?

Ans.—It can; in fact the steam-turbine has solved the problem of superheated steam, owing to the absence of all rubbing parts exposed to the steam. This permits the use of steam of high temperature thus making it possible to realize the advantages of economical operation.

Ques. 798.—Is there not danger of distortion of the turbine cylinder being caused by the very high temperatures to which it is exposed by the use of superheated steam?

Ans.—There have been numerous instances in the past of unequal expansion of the top, and bottom of the cylinder thereby causing the rotating blades to come in contact with the cylinder walls, and be ripped out, but this difficulty has in a great measure been overcome by certain designers of steam-turbines, who have made a special study of the laws of expansion and contraction of metals, and have thus been enabled to make such a distribution of the metal, as to cause an equal expansion of all parts of the cylinder.

Ques. 799.—What effect does the accidental carrying over of water with the steam, have upon the steam-turbine?

Ans.—The sudden presence of a quantity of water with the steam, caused by foaming or priming of the boilers, would cause no more serious results than the slowing down of the turbine during the time necessary to permit the water to be discharged from the exhaust end.

Ques. 800.—What may be said in general of the steam-turbine?

Ans.—It has passed through the experimental stage, and has come to the front, as an efficient power producer, having a bright future before it.

INDEX

A

B

O

P

S

Spotted something that isn't quite right?

Let us know and we'll fix it for future reprints. And as a thank you for your diligence, we'll send you a free Sothis book of your choice.

publisher@sothis.press

This page intentionally left blank

Made in United States
Orlando, FL
05 November 2026

90241899R10215